FEROX AND CHAR

IN THE LOCHS OF SCOTLAND

Perfect
ublishers Ltd

ISBN: 978-1-905399-68-0

Printed by Lightning Source UK.

Published by:

PERFECT PUBLISHERS LTD
23 Maitland Avenue
Cambridge
CB4 1TA
England
www.perfectpublishers.co.uk

FEROX AND CHAR

IN THE LOCHS OF SCOTLAND

AN INQUIRY

BY

R. P. HARDIE

PART II

'The publication of these notes
is perhaps justified
by a long and fairly extensive
experience of lochs in Scotland'

R. P. Hardie (1940)

2011

The Fisheries Society of the
British Isles (FSBI)

IN MEMORIAM

R.P.H. AND G.F.F.

PREFACE

Prior to 1940, when he embarked on the writing of
his extensive account of large Brown Trout and
Arctic Charr in Scottish lochs, R.P. Hardie prepared
a handwritten note outlining his ideas for the work:

> My plan will be this. I shall begin with an
> introductory chapter dealing mainly with the
> classification of the Salmonidae and giving a
> natural history of the two genera in which we are
> specially interested, *Salmo* and *Salvelinus*. Then
> there will be chapters on ferox which is my
> primary subject. Next will follow naturally a
> chapter on the different types of loch. Next there
> will be a chapter on char in general (in Scotland).
> The rest will consist of an account of the more
> interesting particular lochs in Scotland. In this, as
> may be expected, I shall lay special stress on those
> which contain ferox or char. These lochs can
> conveniently be arranged in groups.

In fact, the text which he produced did not follow
this plan exactly, as most of the discussion of Ferox
and Charr themselves was summarised in the
Preface to Part I of *Ferox and Char*. All of the main
text was devoted to an account of the lochs in
chapters on the series of basins from the south of
Scotland, then north up the east coast to the Shin
basin, where Part I ended. Part I of Hardie's study
was published in Edinburgh in 1940 by Oliver and
Boyd.

Unfortunately, Part II of *Ferox and Char* was never published. It was assumed for many years that it had either never been written or that any manuscript he had prepared had been lost or destroyed. In fact, after Hardie's death, many of the relevant papers were kept together and passed by Isabelle Hardie to a Dr Hill for safe keeping. They were subsequently given to Niall Campbell when he started work on Arctic Charr and Niall stored them carefully. Eventually, thanks to Ronald Campbell the files and maps were given to Peter Maitland and, after much sorting and collation, it was found that there was actually a complete handwritten manuscript for Part II, some of it written, apparently, as late as 1940 and 1941.

The present volume, Part II, was prepared from that manuscript. However, the collection of files and letters indicated that Hardie had actually done much of his background research on Arctic Charr in Scotland during the 1920s. Many of the letters written at this time were destroyed but those available showed that he had written around widely to all kinds of people, including landowners, anglers and scientists. Correspondents included such well known names as the naturalist Seton Gordon, the Inspector of Salmon Fisheries, W.J.M. Menzies and J.R. Norman of the British Museum. The latter made it clear that little interest had been taken in Arctic

Charr for many years when he wrote to Hardie in
1927.

> Nothing of any importance has been written on
> these fishes since the publication of Mr. Regan's
> book on 'Freshwater Fishes' in 1911. He dealt with
> them from an evolutionary point of view in his
> presidential address to Section D of the British
> Association at Southampton in 1925 (published in
> 1926), but there is little in this address which is
> not to be found in the above mentioned book.

It was more than 20 years before some of the
material which Hardie had gathered together
appeared in print and it is not clear why there was
such a long delay. He published Part I of *Ferox and
Char*, just in time, for the Second World War was in
full swing by the time it appeared in 1940. The War
is clearly a reason why Part II was not published at
the time, and by 1946, Hardie was dead.

We have tried to follow Hardie's text as closely as
possible and have kept any changes to a minimum.
As the famous phrase goes - the views expressed are
Hardie's and not ours - it is his volume. As in Part I,
he has depended extensively on Murray & Pullar's
(1910) classic *Bathymetric Survey of the Freshwater
Lochs of Scotland* and a high proportion of the text
is taken directly from that phenomenal six-volume
publication. Other writers too are often quoted, for
example Thomas Tod Stoddart, Tate Regan and

Peter Malloch - as he acknowledges in Part I. Again,
as in Part I, Hardie wanders down the occasional
side road in his text, dealing with historical,
geographical, archaeological and other accounts
which are hardly relevant to his main thesis. We
have included all of these and, apart from minor
textual changes for clarity and to avoid the
occasional error, the only extensive input we have
made has been to use the names of the lochs which
are found in the revised 1956 Ordnance Survey One
Inch to One Mile series of maps, rather than those
names in Hardie's manuscript which were often
misspelled, or obscure because of handwriting
problems.

In fact, Hardie's personal contribution of new
information in both Parts of *Ferox and Char* is quite
small and consists mainly of records of the numbers
and weights of Brown Trout, including Ferox, and
occasionally Arctic Charr, found in some of the
lochs. One of the strange things, which is apparent
from the letters and other notes in his files, is that he
did have, mainly from correspondents, original and
valuable information on some populations of Arctic
Charr in Scotland. A good example is a detailed
personal account of fishing for Arctic Charr in Loch
Morie during the mid-19th Century which was
provided in a letter from George Munro of Alness in
1925. It is unclear why he did not include such
valuable new data in his writings.

What is a "ferox"? It is clear that in Part 1 Hardie used the word "ferox" as a colloquial term for a loch dwelling Brown Trout (*Salmo trutta*) which does not migrate to sea and which feeds on other fish. He thus makes a distinction between large trout (often referring to trout of in excess of 15 lbs weight) and ferox, trout which feed on other fish, (but which must also be large for these gape-limited predators to consume such prey). This rather inexact definition has persisted in the scientific literature (Campbell, 1979; Duguid *et al.*, 2006) and here we retain Hardie's original definition. It is interesting (and somewhat disappointing!) to note how little progress we have made in our understanding of this charismatic fish (which is an important component of the diversity of Scottish lochs) in the 70 years since the publication of Part 1. We know that ferox trout are usually long lived, late maturing and reach a much larger body size than other freshwater-dwelling brown trout. We know that in Scotland (as elsewhere) ferox trout are mostly found in large, deep and nutrient poor lochs that also contain Arctic charr or whitefish (*Coregonus* spp.). However all of this was also known to Hardie. Also known to Hardie was that ferox trout typically exhibit head shape differences from non-fish feeders. What was not known at that time was that ferox are frequently (although not always) genetically distinct from other freshwater-resident brown trout within a system (Ferguson & Taggart 1991; Duguid *et al.* 2006).

A useful brief biography of Robert Purves Hardie
(1864-1942) was published shortly after his death by
Smith (1942). After graduating M.A. at both the
University of Edinburgh and then Oxford and
spending a year in continental Europe, R.P. Hardie
was to spend the whole of his working life, 43 years
in all, on the staff of the University of Edinburgh.
He was a noted expert in Greek philosophy and
published several works in this field, but he also had
a gift for metaphysics and pure geometry. He was a
long-standing member of the Edinburgh
Mathematical Society. Initially his post at
Edinburgh was as Assistant to the Professor of
Logic, but later he became Reader in Ancient
Philosophy. His publications were of two kinds.
Professionally he produced several articles on
Aristotle and Plato and helped in the production of
two books on philosophy. Other interests lead to a
pamphlet on the *Tobermory Argosy*, a book on *The
Roads of Mediaeval Lauderdale* and, of course,
Ferox and Char in the Lochs of Scotland, which
Smith said was:

> the fruit of a life-long study of fish and fishing in
> Scottish lochs. Hardie had fished hundreds of
> lochs and kept records for half a century. ...
> Fishing was his favourite recreation: he preferred
> lochs to streams and trout to salmon; 'big game' to
> him meant ferox, for which he long held the
> record.

What then is the value of *Ferox and Char*? Apart from the important work of Albert Gunther, Tate Regan and other fish biologists on the scientific aspects of Arctic Charr in Scotland, and elsewhere in Britain and Ireland, no-one else had previously attempted to put together a comprehensive picture of the distribution of Arctic Charr, and indeed of Ferox Trout, in Scotland. R.P. Hardie must thus be acknowledged as the first person to do so, mainly from the literature and personal contacts, and indeed to have subsequently provided the stimulation for others, notably Kim Friend and those of us who have followed him in the study of these fascinating fishes in Scotland.

Peter S. Maitland
Colin E. Adams

University of Glasgow

ABBREVIATIONS AND SOURCES

These are already given in Part I, but it is useful to repeat them again here in a more modern format.

B.S.   Murray, J. & Pullar, L. 1910. *Bathymetrical Survey of Scottish Fresh Water Lochs*. Edinburgh, Challenger Office.

O.S.A.   *The Statistical Account of Scotland.* 1791-99.

N.S.A.   *The New Statistical Account of Scotland.* 1834-45.

Macfarlane   Macfarlane, W. 1748. *Geographical collections relating to Scotland.* Edinburgh, Constable (published in 1906 for the Scottish History Society).

Regan   Regan, C.T. 1911. *The freshwater fishes of the British Isles.* London, Methuen.

S.G.   Watson-Lyall, J. 1872. *The Sportsman's and Tourist's Guide to the Rivers, Lochs, Moors and Deer-Forest of Scotland.* London, Simkin, Marshall, Hamilton & Kemp.

Stoddart   Stoddart, T.T. 1853. *The angler's companion to the rivers and lochs of Scotland.* Edinburgh, Blackwood.

Other references

Calderwood, W.L. 1941. *Scottish Geographical Journal.* 57, 132.

Campbell, R.N. 1979. Ferox Trout, *Salmo trutta* L., and Charr, *Salvelinus alpinus* (L.), in Scottish lochs. *Journal of Fish Biology*, 14, 1-29.

Collet, L.W. & Johnston, T.N. 1906. On the formation of certain lakes in the Highlands. *Proceedings of the Royal Society of Edinburgh*, 26, 107-115.

Colquhoun, J. 1840. *The Moor and the Loch.* Edinburgh, Blackwood. Two Volumes, and later editions: 1841, 1851, 1878, 1880 ...

Commission on Ancient Monuments of Scotland. 1928. *Ninth Report on the Outer Hebrides, Skye and the Small Isles.*

Davy, H. 1869. *Salmonia; or, Days of Fly Fishing. With some account of the Habits of Fishes belonging to the Genus Salmo.* London, John Murray.

Day, F. 1887. *British and Irish Salmonidae.* London, Williams & Norgate.

Duguid, R.A., Ferguson, A. & Prodöhl, P.A. 2006. Reproductive isolation and genetic differentiation of ferox trout from sympatric brown trout in Loch Awe and Loch Laggan, Scotland. *Journal of Fish Biology*, 69A, 89-114.

Edwards-Moss, J.E. 1888. *A Season in Sutherland.* London, Macmillan.

Ferguson, A. & Taggart, J.B. 1991. Genetic differentiation among the sympatric brown trout (*Salmo trutta*) populations of Lough Melvin, Ireland. *Biological Journal of the Linnean Society*, 43, 221-237.

Groome, F.H. 1882. *Ordinance Gazetter of Scotland.* Edinburgh, Jackson. Six Volumes.

Low, G. 1813. *Fauna Orcadensis; or, the Natural History of the Quadrupeds, Birds, Reptiles and Fishes of Orkney and Shetland.* Edinburgh, Ramsay.

Mackenzie, O.H. 1921. *A Hundred Years in the Highlands.* London, Arnold.

Malloch, P.D. 1910. *Life history of the Salmon, Sea-trout and other Freshwater Fish.* London, Black.

Malmesbury, J.H.H. 1884. *Memoirs of an Ex-minister.* London, Longmans, Green & Co. Two Volumes, 3rd Edition.

Menzies, W.J.M. 1936. *Sea Trout and Trout.* London, Arnold.

Nall, G.N. 1930. *Life of the Sea Trout.* London, Seeley, Service & Co.

Pennant, T. 1777. *British Zoology.* London.

Smith, J.C. 1942. The late R.P. Hardie. *University of Edinburgh Journal*, 10-11, 240-242.

Stoddart, T.T. 1835. *The Art of Angling as Practiced in Scotland.* Edinburgh, Chambers.

Stoddart, T.T. 1886. *An Angler's Rambles and Angling Songs.* Edinburgh, Edmonston & Douglas.

Thompson, W. 1856. *Natural History of Ireland.* London, Bohn. Volume 4.

ACKNOWLEDGEMENTS

We thank the late Niall Campbell and also Ronald Campbell for helping to rescue the handwritten manuscript of this volume. Richard Hunter kindly provided biographical information on Robert Purves Hardie. Financial support for the publication was generously provided by the Fisheries Society of the British Isles.

CONTENTS

I

SHIN, BRORA AND HELMSDALE

Shin (concluded)

In part I of this work, my account of the lochs in the Shin basin was left incomplete. No additional loch is given in the B.S., but several other lochs which run into the River Shin should be mentioned. Firstly, Loch na Fuaralachd and Loch Beannach which run into the River Tirry, which discharges to Loch Shin. Three lochs run directly into the River Shin at Lairg - Loch Tigh na Creige, Loch Dola and Loch Craggie, the last being the highest, about a mile long and holding very good trout. We may also note three considerable lochs on the west side of Loch Shin: Loch Sgeireach, which enters the loch opposite Shinness Lodge, Loch Fuaralaich, and Loch na Caillich which enters the River Shin below Lairg, via the Grudie Burn. One may add two very small lochs about two miles north-east of Overscaig Inn which have good trout averaging nearly two pounds.

We may next include under the name 'Shin' two small basins which debouch, not into the river itself, but into the Dornoch Firth. The first of these enters the sea by a very small stream at Spinningdale, the other by the small River Evelix, at Dornoch. In the latter basin the B.S. gives a description of only one loch, Loch Migdale and also of only one in the former, Loch an Lagain, though it names a second, Loch Laro, at the head of the river, neither being good for fishing. Loch Migdale, on the River Evelix, holds both trout and pike. It is at a height of about 114 ft., is nearly two miles long, and has a maximum depth of 49 ft.

The next basin through which we may pass is that of another small river, the River Fleet. Here again, the B.S. measures only one loch, Loch Buidhe, mentioning again two other lochs, Loch Cracail Mor and Loch Cracail Beag. The first of these lochs is at a height of 527 ft., the other two being higher. Loch Buidhe is 1¼ miles long, and has a sluice by which it can be raised 5 ft. (to induce salmon to enter the loch, though this has not been successful). It is a good trout loch. The maximum depth is only 36 ft., and the mean depth 12 ft. There are at least six other small lochs in the basin.

Before reaching the River Brora, the next considerable river, we have to pass through two more very small basins, those of the Culmally Burn and of the Dunrobin or Golspie Burn. Each contains a considerable loch, but neither was recorded by the B.S. In the former there is Loch Lunndaidh, at a height of 556 ft. and nearly a mile long; it is said to have been stocked recently with Loch Leven trout. In the latter basin, there is Loch Horn which is fully half a mile in length and at a height of 1155 ft. On a tributary of the burn is Loch Caorach at a height of 1250 ft.

Brora

In the Brora basin, only one loch has been measured, [by B.S.] namely Loch Brora itself, but there are about a dozen other considerable lochs, many of them over half a mile in length, e.g. Glas-loch Mor and Gorm-loch Mor. These are all in the old forest of Dirichat near Ben Armine Lodge, at the sources of the two chief tributaries of the River Brora, at the respective heads of Strath na Seilge and Strath Skinsdale.

Loch Brora is 3½ miles long and about 92 ft. above the sea, while its maximum depth is 66 ft. Its mean depth is 22½ ft. The loch is divided by three constrictions which are alluvial in origin. Thus there are four distinct basins in Loch Brora which exceed 30 ft. in depth Proceeding from the foot of the loch, the first, smallest, and shallowest basin has a maximum depth of 31 ft. Passing through the first narrows, where a depth of 7 ft. was observed, one enters the second expansion of the loch which is shallow until Eilean nam Fauleag is passed, the second basin lying north of that island, and having a maximum depth of 43 ft. Passing through the second narrows, where a depth of 9 ft. was found, one enters the third and deepest basin, enclosing the maximum depth of 66 ft., which was recorded near the centre of the basin. Passing through the third and most northerly narrows, in which depths of 9 and 7 ft. were recorded, one enters the fourth and largest basin. This basin is cut into two portions by the slight shoaling of the bottom where the alluvial cone laid down by the Allt Smeorail projects into the loch, the deepest water to the south being 59 ft. and to the north-west 64 ft., while on the shoaling the greatest depth observed was 50 ft.

There is a statement of the fish in the loch in the O.S.A. to the effect that:-

> It abounds with salmon, jar, and other trout of different kinds.

Helmsdale

We come now to the comparatively large basin of the River Helmsdale, with a complicated and rather confusing system of lochs and rivers. To simplify it slightly, there are three main tributaries of the river:

(1) The Bannock Burn, coming more or less from the north, has its source in Loch an Ruathair. This loch is shallow, only 26 ft. deep, and has a sluice on it in order to regulate the spates on the lower river. It holds salmon, trout and char and is the shallowest loch to have char that we have encountered hitherto in our progress round Scotland. The loch, however, has a considerable mean depth of 13½ ft. and a

small drainage area of only three-quarters of a square mile. The loch exceeds 1½ miles in length and is situated at a height of 414 ft. above the sea. The flat-bottomed character of the loch as a whole is shown by the broad spacing of the deep water contour lines in the B.S.

It has pretty surroundings, Creag Sail a' Bhathaich and Meall a' Bhuirich rising from the north-western shore, overshadowed by the peaks of Ben Griam More and Ben Griam Beg (both nearly 2000 ft.), farther distant in the same direction, while to the south-east the Knockfin heights exceed 1400 ft.

(2) The second main tributary comes from the north-north-west and is called the Claggan Burn. It has its source in two lochs, Loch Coire nam Mang and Loch Druim a' Chliabhain, which are reputed to be the best trout lochs in Scotland. Very large takes are reported from them such as 84 trout of 50 lb total weight. The B.S. says that the trout in the larger of the two lochs are large while those in the smaller Loch Coire nam Mang, which runs into the other, are, as we should expect, 'larger'. Ben Griam More lies to the south of Druim a' Chliabhain and Ben Griam Beg to the east.

Loch Druim a' Chliabhain is 1⅔ miles in length. It is 30 ft. lower than Coire nam Mang, and has a maximum depth of 51 ft, which was observed near

the southern end of the loch. Loch Druim a'
Chliabhain is described by the B.S. as:-

> a rock basin in granulitic schists situated where the
> ice became constricted in passing between the
> outliers of Old Red Conglomerate forming the two
> Griam Hills in the east of Sutherland.

Loch Coire nam Mang is similarly a rock basin in
Moine schists. It lies at a height of 800 ft. and has a
maximum depth of 33 ft. and a mean depth of 11½
ft. with a drainage area of only three quarters of a
square mile.

There is a third loch, Loch Arichlinie, outflowing to
the same tributary, but it is only one third of a mile
in length. All three lochs are said to hold char as
well as trout, though the maximum depth of the last
is only 7 ft., the mean depth being 4½ ft. It is
described by the B.S. as:-

> a shallow flat-bottomed basin, apparently in the
> process of being filled up. Geologically it is a drift
> dammed loch on crystalline schists.

(3) The third, and the main, tributary of the River
Helmsdale comes from the west, and has on it a
chain of five lochs which includes the largest lochs
in the basin. Let us enumerate them in the ascending
order.

The lowest, Loch Achnamoine, at a height of 376 ft., is a small shallow loch 2½ miles from Kinbrace Station with a maximum depth of only 8 ft. and a mean depth of 4½ ft. It serves as one of the salmon beats on the river. Then we ascend first to Loch Badanloch at 394 ft. - a large loch 1¾ miles long and nearly a mile wide, with a maximum depth of 42 ft. and a mean of 17⅓ ft. The loch holds salmon, ferox, ordinary trout and char.

At the same level is Loch nan Clar, making with the preceding, a continuous sheet of water 3¼ miles in length, the passage between the lochs being about 200 yards long with a depth of 5 ft. The maximum depth of Loch nan Clar is 32 ft. and the mean 13½ ft. The B.S. notes:-

> The surroundings of the loch are very fine, several lofty peaks being visible: Morven (in Caithness) to the south-east, Ben Griam More to the north-east, Ben Armine and Ben Klibreck to the south-west and Ben Hope and Ben Loyal to the north-west.

The fourth loch in the chain is Loch nan Cuinne, known locally as Loch Rimsdale, at a height of 395 ft. It is three miles in length and is only a half foot above Loch nan Clar, with which it is connected by a stream 100 yards in length. The maximum depth of Loch nan Cuinne is 28 ft. and the mean depth is 12½ ft.

Finally, there is Loch Truderscaig, running into the
last loch from a height of 426 ft. It is triangular and
the outflowing stream, the Allt Tarsuinn, leaves the
loch at the apex of the triangle and flows into Loch
nan Cuinne. As we should expect from its
shallowness and flat-bottomed character, it is a very
good fishing loch. The maximum depth is 12 ft. and
the mean rather less than 6 ft.

To this group of five lochs we may append Loch
Alltain Fhearna which falls into Loch Badanloch
from a height of 432 ft. It is nearly a mile long and
the maximum depth is 36 ft., the mean being 14⅓ ft.
It is a good fishing loch, the average trout being
about a third of a pound, but char are not mentioned.

There is an account of the lochs in the district in the
N.S.A.:-

> The upper district of Kildonan is remarkable for
> the number and the size of its lakes. Loch-na-Cue
> is one of the largest of them. It has char and other
> varieties of trout. Loch Leam-na-Claven has trout
> of different varieties of the largest size of any lake
> in the district. There are also a great many char in
> its waters, but they are of a small size. Loch
> Badan-loch and Loch-in-Ruar are also large lakes
> and abound in trout and char.

Smaller lochs in the district may be mentioned.

The smallish Loch Gaineimh runs into Loch nan Cuinne from a height of 725 ft. and is fully half a mile long, which is said to have been stocked in 1925. There are also some very small lochs near the head of Loch na Cuinne.

Two lochs near Loch an Ruathair should be noted, first a considerable Loch Culaidh near its head which presumably runs into it, and has good trout of perhaps three quarters of a pound. Secondly, a very small and rather peaty loch which is near the foot and has some biggish trout of 2-3 pounds, Loch Ascaig, nearly half a mile in length, is on a very small burn running into the River Helmsdale near Borrobol.

II

WICK, THURSO, FORSS AND HALLADALE

Ten miles after we have passed Helmsdale and the Ord of Caithness in our progress up the east coast of Scotland, we come to two rivers which join each other within three furlongs of the rock-bound coast and together enter the North Sea - the Berriedale Water and the Langwell Water. We may conveniently regard the latter as a branch of the former, which is the longer, and presumably the larger, river.

The basin of the Langwell Water, which contains the deer-forest of that name, yields only one loch, and it is very small and very high (1398 ft.), namely Loch Scalabsdale, the source of the Langwell Water, and between three and four miles from Morven (2313 ft.), the highest hill in Caithness. The whole basin of the Berriedale Water contains only two other lochs and these are still smaller.

The next basin we enter, that of the small Dunbeath Water, contains two lochs which are considerable in size, Loch Dubh and Loch Breac. The former is half a mile long and at a height of 750 ft. The loch in which the river rises is smaller - Loch Braigh na h-Aibhne - and sits at a height of 980 ft.

Wick

From this point of the coast onward there is no loch that we need to mention until we reach the basin of the Wick River, one of the three chief rivers in the county, in which three lochs have been measured; two others are mentioned by the B.S.

We may begin with Loch Watten, which is the lowest loch in the Wick basin and the largest in Caithness. Its height above the sea is 55 ft. and it is three miles in length. Its maximum depth is 12 ft., no fewer than 36 soundings being taken at this depth in the south-eastern part of the loch. The mean depth is 8½ ft., i.e. 70 per cent of the maximum, and more than half of the lake floor is covered by more than 10 ft. of water. Thus Loch Watten may be described as a large, shallow, flat-bottomed loch, the water shoaling gradually towards the head. According to the miller at Watten, the wind sometimes perceptibly affects the level of the water and after an easterly wind had been blowing strongly for some time it was impossible for him to work the mill, the water being driven before the wind and piling up at the north-west end. We happen to have an early record, from the O.S.A. at the beginning of the eighteenth century, which refers to Loch Watten:-

> The lochs of Caithness abound with trouts of diverse kinds, and Eels of a huge length and bigness. In Watten there is a fish as big as a

salmon but few or none have been taken since the
English were there, who used to fish them.

The 'English' referred to are almost certainly the
Cromwellian army which occupied Scotland even to
the extreme north and are mentioned in the records
of the presbyteries and parishes of Caithness. We
have more than once come across Richard Franck,
who was a captain or trooper in the English army of
occupation, and we know that he was a skilful fly
fisher for salmon. Loch Watten is not a loch that we
should naturally class as a ferox loch, but it is fairly
spacious, being large and moderately deep (mean
depth: 8½ ft.), and may have contained trout of, say,
6 lb., in the second half of the seventeenth century.
We can readily believe that the English invaders,
being accustomed to pike, were much more expert
than the natives of Caithness at trolling or spinning
for large trout, and had much better rods and other
appliances for the purpose.

The second lowest loch in the Wick basin is the
small Loch Scarmclate, to the north-west of Loch
Watten and surrounded by cultivated land -
reclaimed perhaps from Loch Watten, as the
margins of the loch are swampy and reedy. The
floor of Loch Scarmclate, which is nearly a mile in
length, is practically uniform in depth, namely 5 ft.
The deposits covering the lake floor are brown
muds, except over a small area to the south of the
island, where the deposit is white and calcareous,

and where it was a regular practice to dredge the loch and to use the mud for marling the land. The geological description of the loch is 'lying in boulder clay in a wide open valley in Caithness flagstones' while that of Loch Watten is 'shallow loch ponded in boulder clay in a wide open valley.' The Caithness flagstones floor a large area of the loch along the northern shore.

The next highest loch in the basin is Loch Hempriggs, 155 ft. above the sea. The level of the loch is regulated by a sluice, and the cracked peaty soil at the margin, with boulders covered by a luxuriant growth of *Fontinalis*, indicated, at the time of the B.S., a very recent fall of 2 ft. from a former long-maintained level. The present maximum depth is 8 ft., with a mean depth of 5¼ ft. The loch, which is described geologically as 'a shallow loch on boulder clay lying on Caithness flagstones' lies about 2 miles south-west of Wick, and within a mile of the North Sea. However, the outflowing stream pursues a long and devious course before joining the Wick River on its way to the sea. Like the other lochs in this locality, it is a shallow flat-bottomed basin, over two-thirds of the lake floor being covered by more than 5 ft. of water; the mean depth of the entire loch is about 65 per cent of the maximum depth.

The next highest loch in the county, the Loch of Yarrows, runs into Loch Hempriggs, but it has not been measured. Its height is 301 ft., but it seems to contain only small trout. Lastly, Loch Toftingall (235 ft.) is at the head of the Burn of Acharole, one of the two burns which, with the outflow from Loch Watten, unite to form the Wick River. At the heads of the two burns are two groups of very small dubh lochs. Loch Toftingall is about one third of a mile long, and is said to have trout of good quality.

To the lochs in the Wick basin we may append the little Loch Sarclet. It is very near Ulbster on the coast and a few miles south of Wick at a height of 130 ft. and seems to have its outflow to the sea by a separate stream. The S.G. says it contains trout of fine quality.

We must now interpolate three lochs, each of which is in a separate basin.

First, there is the Loch of Wester, the source of the Water of Wester, which, after a course of a little more than a mile, enters the North Sea about four miles north of Wick. The loch is a mile in length and only 6 ft. above the sea. When the tide is out there is a perceptible current in the river. The loch is very shallow, the maximum depth of 3 ft. occurring at various places throughout the loch, which contains sea trout and small brown trout. On the Burn of Lyth, which enters the loch, there is no other loch, except a small mill dam near the head.

Then, rounding the point at John O' Groats, we come next to Loch Heilen, which at one time was supposed to be the source of the Burn of Lyth, but is now known to have its outflow by the Burn of Instack, which runs to the east of St John's Loch. Loch Heilen is a mile and a half long and is at a height of 113 ft. The maximum depth is 5 ft. and the mean depth half of that.

We come lastly to St John's Loch, which is similar to Loch Heilen and enters the sea by the Burn of Dunnet. It is a shallow, flat-bottomed basin, about a mile in diameter, at a height of 71 ft., and has a maximum depth of 5 ft. and a mean depth of 4½ ft. It holds good trout averaging nearly a pound.

Thurso

The next basin with which we have to deal is that of the River Thurso, the second of the three comparatively 'large' Caithness rivers. Only one loch is measured here by the B.S. - Loch More, a well known salmon loch. The loch is described geologically as a remnant of a larger loch through which the Thurso River flowed. It lies in a hollow of drift material, but it seems highly probable that the drift conceals a rock basin, as the river at Dirlot is excavating a rock gorge through the local members of the Caithness flagstones into the underlying schists. Like most of the Caithness lochs, Loch More is a shallow, flat-bottomed loch. Most of the soundings were taken in depths of 5-6 ft. The deepest cast, in 7 ft., was taken about 100 yards from where the River Thurso flows out, while along the southern and northern shores, the bottom is silted up by the deposition of material laid by the River Thurso and the Sleach Water. The area of the lake floor covered by less than 5 ft. of water is about 92 acres or 52 per cent of the total area of the loch.

The loch was surveyed on October 2, 1902, and the height of the surface was found to be 381 ft. and the maximum depth 7 ft. The mean depth was 4 ft., while the surface area was about 177 acres. In the present century, the loch level has been raised with the object of improving the salmon fishing in the River Thurso. But the salmon fishing in the loch has been injured. On the other hand, the trout fishing has much improved, at least for the time being by the extension of the loch, as is shown by the photograph in P.D. Malloch's *Life History of the Salmon, Sea Trout and other Freshwater Fish* - 'Loch More trout averaging 2 lb., showing condition greatly improved by extended feeding area of the loch, July, 1908.'

The basin of the Thurso contains many other lochs besides Loch More, some of which are worthy of mention. We may group them as follows under their sub-basins and tributaries.

(1) The stream in Strath Beag, the 'Little River', which joins the River Thurso on the right bank

between Dirlot and Westervale has four 'considerable' lochs on it. Loch Ruard, which is more than a mile long and holds many trout averaging half a pound. There is a small loch which runs into Loch Ruard and may have good trout. The other two lochs are Loch Rangag and Loch Stemster which 'is famous for its marl, trout of 2 or 3 lb. being often got in it.'

(2) The Sleach Water, which enters Loch More from the west. On this tributary there is a group of fifteen or more lochs (the largest of them about three-quarters of a mile long) which include Loch Eileanach and Caol Loch, which has good trout.

(3) There is the River Thurso, with Loch a' Mhuillinn and its tributary Loch nam Fear, entering Loch More from the south, with the Allt Backlass entering just to the east. In the latter sub-basin there is Loch Sand with a tributary loch, Loch Thulachan. On the Glut Water, near the source of the River Thurso, there are two dubh lochans which are said to hold large trout.

(4) Lastly there are two lochs which run directly into the River Thurso itself, below Loch More, firstly Loch a' Cherigal and further downstream, Loch Meadie. The latter is more than three quarters of a mile long, while the S.G. says that the former has yielded trout of 5 or 6 lbs.

Thus the whole basin contains at least twenty significant lochs in addition to Loch More.

Forss

We now proceed to discuss the lochs in the third 'large' river in Caithness, that of the River Forss. Two lochs have been measured by the B.S. Loch Calder, which is on a small tributary of the River Forss, the Allt an Ghuinne, and lies about five miles south-west of Thurso, which is five or six miles south-east of where the River Forss enters the sea. The other measured loch, Loch Shurrery is nearer the source of the River Forss. Loch Calder differs from other Caithness lochs we have described, because of its greater depth, which has a maximum of 85 ft. The height of the surface is 205 ft., but it is affected by a sluice at Achavarn on the east shore, operated by both the South Calder Mill and the water works at Thurso. The loch is irregular in outline and rather peculiar in conformation. In the wide portion of the loch, off the western shore, there is an island situated on a large bank surrounded by deeper water, and the narrow southern portion is so shallow that one must proceed three-quarters of a mile from the southern end before meeting with depths exceeding 11 ft. The deep basin is contained in the eastern half of the wide northern portion of the loch, the deepest sounding of 85 ft. having been taken about half a mile from the northern shore and a quarter of a mile from the eastern shore. Here there is a basin about a mile in length and exceeding 30 ft. in depth, the basin of 50-foot or greater being nearly three-quarters of a mile in length, and distant about a quarter of a mile from the northern shore. The loch as a whole is comparatively shallow, since 72 per cent of the lake floor is covered by less than 25 ft. of water.

There is reference to the fish in the loch in the O.S.A.:-

> In the lake of Cathel [this seems to have been a sheiling near the loch] there are trout which are nowhere else in the country, of a reddish beautiful colour, a pretty shape, very fat, and most pleasant eating. I suspect they are that kind of fish which naturalists call fresh-water herring.

The char in the loch are usually described as "small" char. There are also plenty of trout, averaging about three to the pound and, it is said, some large trout.

The other measured loch, Loch Shurrery, the source of the river, at a height of 321 ft. is eight miles from Thurso and seven miles from Reay. It is a shallow flat-bottomed loch, 1¼ miles long, the majority of the soundings having been taken at depths of 5 and 6 ft., while three soundings were taken at the maximum depth of 7 ft. The loch is peaty and unattractive, but holds a few salmon.

Though only two lochs were measured in the basin of the Forss, and only another two are mentioned by the B.S., one or two other lochs in the basin are worthy of naming at least.

Further up the valley there are Loch Scye and Little Loch Scye, which have good trout averaging three quarters of a pound.

Then, on the Cnocglas Water, as the river above Loch Shurrery is called, there is Loch Caluim, about half a mile square, at a height of 435 ft., with good trout of nearly half a pound, and a small tributary loch, Lochan na Saighe Glaise, with larger fish; and nearby Loch Losgann (trout averaging three quarters of a pound) and Loch Chiteadh, which enter the Cnocglas Water opposite Loch Caluim.

Further up the Cnocglas Water there are three fair-sized lochs which have small trout, Loch Torr na Ceardaich, Loch Ealach Beag and Loch Tuim Ghlais. At the sources of the latter are Loch a' Chleirich and Lochan Dubh Cul na Beinne, where there are said to be larger trout.

Finally, running into the head of Loch Calder, there is the uninteresting Loch Olginey, which is shallow and very peaty.

To the west of the River Forss is a small tributary which runs through Dounreay. On it there are two considerable lochs, Lochs Thormaid and Loch Saorach. The larger, Loch Saorach, is fully half a mile long.

We now pass near three very small basins, those of the Achvarasdal, Reay and Sandside Burns. Loch nan Clachan Geala is on a branch of the first of these and is said to have good trout and there is a loch with a similar name, Loch nan Clach Geala, on a branch of the third.

Halladale

When we pass from Caithness to Sutherland, we enter the basin of the River Halladale, a fairly well-known salmon river. The lochs in it are not given by the B.S., presumably because they are not deep and are interesting only to trout fishers Some of them, however, are very good trout lochs and deserve some recognition.

Near Forsinard, at the head of the valley, there are Lochs Sletill, Leir, Talaheel and nan Clach Geala in which the trout average about three-quarters of a pound. P.D. Malloch, in his *Life History of the Salmon, Sea Trout and other Freshwater Fish,* has a photo of 12 Sletill trout, averaging 1 lb., presumably the best 12 of a day's catch (July, 1900).

Then towards the foot of the basin, on the left bank, Loch na Caorach, Loch na Seilge, Loch Akran and

Caol-loch are well known for their trout; also, on the right bank, Loch Achridigill, Loch na h-Eaglaise Mor, Loch na h-Eaglaise Beg and Lochan Coulbackie.

We come next to the basin of a small salmon river, the River Strathy, which also contains several lochs that are not in the B.S. Loch Strathy at its source (646 ft.) is rather remote, but fairly well known as a trout loch. At the source of another branch, Loch na Saobhaidhe (705 ft. high) is almost as good as Loch Strathy. P.D. Malloch illustrates trout from Loch Balligiull in the same basin. At the sources of other tributaries are found Loch Suidhe and Loch nan Clach.

The next basins we come to are those of three burns which enter the sea at Armadale Bay, Kirtomy Bay and Swordly Bay. In the first basin there are two considerable lochs on Armadale Burn: Loch Buidhe Mor and running into it Loch Buidhe Beag. In the third basin, one of the many lochs called Loch Meadie, a mile and a half long, on the Swordly Burn, but holding only small trout and none of them measured.

III

ORKNEY AND SHETLAND

We have seen that the main characteristic of the north of Scotland is that the rivers are short, and that the number of river basins is in consequence correspondingly multiplied. When we leave the mainland and move on to the major island groups (Orkney, Shetland, North Uist and Lewis) we find that this feature becomes more conspicuous, especially in North Uist. There, the land is broken up into a multiplicity of islands and peninsulas, and the lochs too become very numerous. Only a drastic process of selection can cope with the situation.

Orkney

George Low, in his *Fauna Orcadensis*, includes the char, and describes it thus:-

> Very infrequent with us, though sometimes seen in the Loch of Stenness. I have not seen above a couple of specimens. Perhaps we know not the method of catching them, or the season may differ here from other places where they are found. However this is, they are not much known. It is a beautiful red in the belly which distinguishes them from the other trout of the loch, and it is a sufficient mark to strike the most inattentive observer.

The work was not published until 1813, but is supposed to have been written between 1777 (the date of Pennant's *British Zoology*) and 1795.

Low divides the genus *Salmo* into five species. (1) The Salmon. (2) The Bull Trout ('not so large as the true trout, here called Burn Trout'). (3) The Trout. (4) The Parr. (5) The Char. Under 3, we have a note 'I have heard of a species of trout (but what species I know not) weighing twenty-three to thirty-six or more pounds; but these, I daresay, are uncommon.'

To judge the value of Low's evidence about Loch Stenness we must have before us a description of the loch, which we can obtain from the B.S.

The Survey points out that Sir Walter Scott refers to 'Loch Stenness' as the combination of the loch to which we now give the name, and the larger Loch Harray which is immediately to the north of it, connected with it, and on the same level.

Loch Stenness (in this narrow sense) is nearly four miles long, and enters the sea at the head of the Bay of Ireland, where the road from Kirkwall to Stromness crosses by the Bridge of Waith. At the eastern extremity of the loch, it links with Loch Harray:-

> though the channel between Loch Stenness and the sea is broad, the level of the loch is but little

affected by the tides, which indicates that the bar
is but little below ordinary high water level. The
loch is flat-bottomed, and has a mean depth of
10½ ft. and a maximum depth of 17 ft.(B.S.).

On the whole, one would be inclined to say that
Low's statement about char in Loch 'Stenness' in
1776 is probably true. There were char, as we have
seen in Part I, in Loch Leven, as late as 1837. Both
are shallow lochs. Loch Stenness is only 17 ft. deep
and Loch Harray is even less – 14 ft., whereas Loch
Leven, at the time under consideration, had a
maximum depth of over 90 ft. Thus the last named
had an area – though not very extensive – of cold
water at the bottom, which is of great importance to
char-life, and would be almost absent in Loch
Stenness.

On August 9, 1903, the temperature at the surface of
Loch Stenness when the survey was made was 60.2°
F., perhaps too high for char in so shallow a loch. In
Loch Harray, the conditions were more favourable
for the fish, the temperature being 55° F., both at the
surface and at the bottom.

Probably the slight salinity of the water would affect
the question of char. As we shall see later, there is a
char loch in North Uist which the tide enters, though
only to a slight extent. While in 'Stenness', the outer
loch, so to call it, marine algae grow throughout and
the fauna is marine; in Loch Harray, no sea weeds

are seen on the east side of the bridge and the water
of the loch is fresh to the taste.

The best commentary on Low's statement that there
are trout in Loch Stenness 'weighing twenty-three to
thirty-six or more pounds' is perhaps the fact that a
trout of 29 lbs. was got in the loch in 1889. Of this,
Mr W.J.M. Menzies remarks 'In that curious tidal
loch with the mixed population of trout, it may have
really been a true sea trout although described at the
time as a 'slob trout'.' This however seems unlikely
and P.D. Malloch tells us in his *Life History of the
Salmon, Sea Trout and other Freshwater Fish* that:-

> Loch Stenness, in which the water is always
> brackish, contains many brown trout which, owing
> to the splendid feeding ground, ran to a large size.
> I have noticed many trout from 10 to 20 lbs. in
> weight rising in Loch Stenness, but these large
> ones are very difficult to catch. Having an
> opportunity to feed all the year, they are always in
> good condition.

Leaving then the 18th century and Loch Stenness, we
find that there is a loch in Orkney that was known in
the 19th century, and up to a comparatively recent
date, to contain char. It is on the mountainous island
of Hoy and is known by different names, but
presumably is the Heldale Water of Bartholomew's
1894 map. The loch is more than one mile long and
seems to be at a height of 193 ft.

Mr Regan calls it Loch Hellyal and gives the following account of it:-

> In 1862, Dr Traill captured two of these char, fishing with a worm. These are both males, 7 to 8 inches long, and are now in the British Museum. In the rather slender body, the blunt snout and the rounded lower jaw included within the upper, the species resembles the Rannoch char; but there the resemblance ceases, for the head is short, the eye small, the interorbital region convex and rather broad, and the fins small, moreover the caudal peduncle is more slender than in the Struan, twice as long as deep.

> During the last few years (1911), Mr William Cowan has very kindly made several attempts on my behalf to get more examples of this interesting fish, but without success. Mr P. Middlemore, who owns the lake, has also made unsuccessful attempts to catch some char; none has been captured since he has been proprietor, and he believes they are extinct. Char are not known from any other lake in the Orkneys.

On these data Mr Regan has recognised this fish as a distinct species (*S. inframundus* Regan).

The only loch on Hoy which is given in the B.S. is Hoglinns Water, which is a rather interesting loch. It lies among heather-covered hills of about 600 ft. in height. It is a little more than a third of a mile long, and a fifth of a mile broad, and is by far the deepest loch in Orkney, having a maximum depth of 57 ft., somewhat west of the centre. It is a simple basin, deeper towards the west end, and has a mean depth of 26 ft. The drainage area measures scarcely half a square mile. The outflow is westwards by the Hoglinns Burn.

It seems conceivable that this really was the loch that held, and perhaps still holds, char.

Shetland

In Shetland, our interest naturally centres on the Loch of Girlsta (or the Black of Girlsta, as it is sometimes called, presumably from its depth). As the B.S. informs us, of the thirty-one lochs that were sounded by the survey, the Loch of Girlsta is actually the deepest, having a maximum depth of 74 ft., compared with the 73 ft. of Clings Water which comes next and is also second in point of mean depth – 26½ ft. as compared with the 31 ft. of the Loch of Girlsta.

The Shetland Islands are very different in their physical features from the neighbouring group of the Orkneys. In place of the tame undulating surface of Orkney, the Shetlands, though not higher, are more rugged and more varied. High rocky ridges are separated by deep valleys, both running north and south. The more varied surface gives rise to a greater diversity in the lochs. Though many are shallow, there is not the unvarying flat-bottomed character of the Orkney lochs, and some are relatively deep. Though there are many basins in which there are numerous lochs, it never happened that it was possible to survey more than two in the same basin, and, in so many cases was there only one loch in the basin sounded, that the thirty-one lochs surveyed by the B.S. occupy twenty-four separate basins.

The Loch of Cliff (in Unst, the most northerly of the islands), with the most extensive drainage system in Shetland, drains an area of only 8½ square miles. The longest loch on Shetland is Loch Strom on the mainland. It has also the greatest superficial area, a little over half a square mile. The largest body of water, however, is the Loch of Girlsta, which, inferior both in length and in area to the two Lochs, Strom and Cliff, has nearly three times the volume of water of any other loch in Shetland.

One is at once attracted to the fact that the Loch of Girlsta is the deepest (74 ft.) loch in Shetland and also the only loch that has char. The loch is 1½ miles long and has a mean depth of 31 ft., which is a high percentage of the maximum depth, and the greatest in Shetland. Only small burns enter the loch, the largest being the Bretta Burn (with four small lochs on it), and the drainage area is barely two square miles. The long axis of the loch runs

north and south. The outflow, controlled by a dam
and sluice, is by a mill lade, one-third of a mile long
to the Mill of Girlsta and then into Wadbister Voe.
The shores are desolate heather-covered hills, rising
on east and west, higher on the west. There is a
broad beach of small grey stones on the east and
west and a sandy beach at the north end. The island
in the loch is low and heather covered. Rock is
exposed on the island and at several spots on the
east shore. Near the outflow there is rock in vertical
strata worn to the level of the beach. The basin is
very simple, with approximately even slopes on all
sides to the deepest water in the middle. The
temperature on July 8, 1903, only varied by 0.3°F.
from surface to bottom, the surface temperature
being 54.1°F. and, at 75 ft., 53.8°F.

Besides char, 'which are occasionally caught', there
are large trout in the loch, perhaps ferox:-

> The fish in the loch average 1 lb. each, and trout
> up to 7 lb. have been caught. The most deadly lure
> is the natural minnow and parr tail or small trout
> should be used for large fish. Large silver or blue
> phantoms also do well. (S.G.).

Of the Shetland char (*Salvelinus gracillimus*) Tate
Regan, the discoverer, says:-

> It is also a very distinct form. It is found in the
> Loch of Girlsta, Tingwall, whence some have been

sent to me by Mr J.S. Tulloch, who tells me that Girlsta is the only char loch in the Shetlands. This species has the body more elongate than any other char, the greatest depth being contained from five and a half to six and a half times in the length of the fish, measured to the base of the caudal fin; the form of the snout, blunt and somewhat truncated is also peculiar; the lower jaw is not, or scarcely, shorter than the upper, the eye is rather large and the fins are well developed. The colour of the back and sides, with the dorsal and caudal fins, is bluish grey, that of the belly silvery or orange; orange spots are present on the sides. This is a small species attaining a length of eight inches.

I may note in conclusion, that the loch is only 87 ft. above the sea, but it is far north (60° 15'N).

I may add a note or two about other lochs in Shetland, none of which, however, is very interesting. The best known lochs on Shetland are, as we should expect, those nearest the present capital of the islands – the Loch of Tingwall and the Loch of Clickimin.

Loch Tingwall is a small tidal loch lying close to the west of the town of Lerwick. A channel to the sea is cut through a low bar of gravel; on the west rises a steep hill. The east and south shores are strewn with boulders, and there are also many in the loch. A broch or dun stands on a promontory strewn with stones. Loch Tingwall is a fairly large loch of oval

form. It occupies, with the Loch of Asta, a small
loch into which it runs, a narrow valley, and is just
over a mile long, its maximum depth being 60 ft.
and its mean depth about 19 ft. The valley in which
the two lochs lie runs along the mainland of the
island, nearly north to south. The whole west side of
Loch Tingwall is cultivated.

Loch Tingwall is divided into two nearly equal
portions by a constriction, where it is also very
shallow, the depth in the centre of the narrow being
only 9 ft. It is still shallow in the channels on either
side of the island north of the narrows, where the
depths are only 2 and 3 ft. The northern basin is the
shallower. It is almost flat-bottomed over the greater
part of the area, with depths of from 7 to 9 ft. In the
centre is a very small and abrupt depression, in
which there is a depth of 40 ft. The southern basin is
of a different character, the sides slope fairly
regularly to the centre, where there is a maximum
depth of 60 ft. The mean depth is about 19 ft. No
large streams enter the loch; its outflow is to the
south, by a stream about 100 yards long, into the
Loch of Asta.

The Loch of Asta is barely half a mile long, and 10
ft. in maximum depth, with a mean depth of
between 5 and 6 ft. It is said that only high tides
enter, and there are no fish in the loch. There is rock
close to the surface on the west side and the burn
flows among stones, with rock in the channel a few

yards from the loch. This loch is narrow and elongate from north to south, with a length of half a mile. It is very shallow, the greater part less than 6 ft. deep, with a single sounding of 13 ft. near the northern end. The level is barely 2 ft. lower than the Loch of Tingwall, 26½ ft. above sea level, nearly identical with that determined by the O.S. in 1876.

Both lochs held good trout averaging 1 lb. 60 years ago, with a chance of getting a sea trout in the lower loch which enters the sea near Scalloway, the old capital of the island.

In conclusion, I may describe some of the chief river basins which hold more than one loch.

Loch Spiggie is the only moderately large loch in the southern part of the mainland of Shetland. It is approximately oval in shape, and its long axis runs nearly north and south. The surrounding country is low and there are a number of farms on the shores of the loch. In length it is fourth in the lochs of Shetland, measuring 1⅓ miles in a straight line between the extreme points.

The loch forms a simple basin, very shallow near the north end. The maximum depth is 41 ft. The outflow is through a narrow bar, the Beach of Spiggie. The principal streams entering the loch are that from the Loch of Brow, the Burn of Hillwell, near the south end, and the Burn of Scousburgh in the north. The surface is only 4 ft. above the sea.

The small Loch of Brow is half a mile long, and is very shallow, with a maximum depth of 6 ft. and a mean depth of 2½ ft. It is at the same height as the Loch of Spiggie and the outflow to that loch is through a flat boggy meadow.

Both lochs have good trout, the former averaging about a pound.

We may now turn to the neighbourhood of Walls, a district which contains many lochs, of which we may select the Upper Loch of Brouster as more or less basal. From it a short stream runs through the Bridge of Walls into the Loch of Brouster, a sea loch and a branch of the Voe of Bowland. The Upper Loch of Brouster is the lowest of an extensive chain of lochs extending from near Sandness (the western extremity of the mainland) to the Voe of Browland. The level of this loch was not found, but it was stated that the water might fall a little lower, and might rise to 5 or 6 ft. higher than at the date of the survey (July 20, 1903).

The next higher loch in the basin is the Loch of Voxterby, which runs by the Braca Burn into the Upper Loch of Brouster at its north end. But it has not been surveyed, though it seems large – almost a mile long. Two burns run into the Loch of Voxterby.

(1) The Gibbie Law's Burn from a higher loch, Burga Water, which descends in turn from a still higher loch, Mousavord Loch. The former was surveyed but not the latter, which is smaller. (2) Another burn comes from the considerable Lunga Water, which was not, however, surveyed, but is about half a mile long.

The length of Burga Water is fully half a mile, and the maximum depth 31 ft. is in the south end, the mean depth being nearly 13 ft. On the date of the survey (July 21, 1903) the height of the loch was 116 ft., the water being low at the time and liable to rise 3 ft. higher. The chief sources of the loch are the Burn of Cattykismires from the west and the burn from Mousavord Loch. Gibbie Law's Burn flows from the east corner 1½ miles south-eastward into Voxterby Loch.

Lunga Water seems to be at a height of 175 ft. The burn from it, which runs into the Loch of Voxterby, passes through the Loch of Whitebrigs on its way thither.

A little to the west of the extensive Brouster basin a small basin opens to the sea, with two lochs on it, the Loch of Kirkigarth and Bardister Loch, an upper and lower loch, while a considerable distance to the north-east is the inlet called The Vadills, famous for sea trout, with a burn into the sea which has on it no fewer than six lochs, the lowest being the small

Loch Culeryin, well known for sea trout, and
including Grass Water, a shallow weedy loch, two
thirds of a mile in length and Hulma Water which
larger; also on three tributaries, Heilia Water, Sand
Water and Turdale Water – a considerable basin. A
little further east is the basin which terminates in
Clings Water and contains another (lower) loch,
Loch Setter, 44 ft. high. Clings Water, it will be
remembered, is the second deepest loch in the
island. It receives only surface drainage and may
hold char.

> The portion of the mainland called North Roe is
> like North Uist or Benbecula. The crack
> containing the lochs is situated to the north of
> Ronas Hill and towards the west coast. Seen from
> the top of Black Butten, one of the summits of
> Ronas Hill, the scene is like that viewed from Lee
> in Uist, but of less extent. The lochs are seen
> thickly studded over a stretch of rough dark
> moorland some four miles long by three miles
> broad. The moor undulates a good deal between
> 350 and 500 ft., and most of the larger lochs are
> near the latter level. (B.S.).

We may describe two of these lochs in this strange
region of North Roe, which are both in the same
basin.

Roer Water is the largest of the lochs of the North
Roe. It is close to the foot of Ronas Hill. The O.S.
gave its height as 349 ft. in 1876. It is nearly two

thirds of a mile in length, with a maximum depth of 32 ft. and a mean depth of 10 ft., and all except two very small holes less than 16 ft. deep. The drainage area extends to 2¼ square miles and includes many small lochs.

Clubbi Shuns is the smallest in volume of the lochs measured on the North Roe, and runs into Roer Water. The drainage area of rather less than one square mile includes very many small lochs and the large Tonga Water, which were not surveyed. The outflow from Tonga Water is via the burn passing through Clubbi Shuns into Roer Water. The level of Clubbi Shuns would be over 350 ft.

In conclusion, I may give a few words to Unst, the most northerly of the islands of the Shetland group. There are four lochs on this island, which was supposed to be the *Ultima Thule* of the Romans and whose latitude is 60°N.

The largest of these lochs is the Loch of Cliff which is just a little shorter than the Loch of Strom, being 2⅓ miles between its extremities. It runs north and south in the long valley which occupies the whole centre part of Unst from north to south. The hills to the west are high, attaining 558 ft. in Libbers Hill, heather clad in the higher parts with pasture below. Near the northern end of the loch, a long arm runs to the south-east, which is filled with weeds, and at the south end the loch is also weedy. A flat area bars the northern end of the loch and through this the Burn of Burrafirth cuts a zigzag course into Burra Firth. The drainage area of 8½ square miles exceeds any other loch in Shetland and includes the Loch of Watlee, a fairly large loch lying three miles to the south. On August 4, 1904, the surface level of the Loch of Cliff was 6 ft. above the sea.

The B.S. has measured a second loch on Unst which, however, is in a different basin. This is the Loch of Snarravoe in the southern part of Unst. It is a narrowly triangular loch, broadest in the south, its axis running north-east to south-west. The largest burn is that coming in at the north-east end from the Loch of Stourhoull, half a mile up the valley. The Burn of Snarravoe flows out at the south-west corner and winds a quarter of a mile north-westwards into Snarra Voe. The loch is over half a mile long. The maximum depth is 29 ft. At the date of the survey, the elevation of the loch was 5⅓ ft. (August 6, 1903).

A small Loch of Belmont, in the extreme south-west end of Unst, is said by the S.G. to have 'excellent trout – 2 or 3 lb. each', which, however, are shy. It is six miles from the Loch of Cliff, not three miles as the S.G. says.

IV

NAVER, BORGIE, KINLOCH, HOPE AND DURNESS

Naver

We pass now to the River Naver, an important salmon river which rises in Loch Naver and enters the sea at Bettyhill.

The river has three main branches.

(1) The main source of the River Naver is the loch of the same name, which is more than six miles in length, and at a height of 247 ft. It has a maximum depth of 108 ft. and a mean depth of 39 ft.

(2) A secondary source is Loch Coire, which is more than three miles long, at a height of 570 ft., with a maximum depth of 151 ft. and a mean depth of 39 ft.

(3) A third source is Loch Meadie, which is at a height of 488 ft. and 3⅓ miles in length; its maximum depth is 63 ft. and its mean depth 20½ ft.

The second and third of these lochs are connected with the main river by the River Mallart and River Mudale respectively, the former entering it a short distance below the outlet of Loch Naver, while the latter enters the loch at its head.

Besides Loch Naver, these two lochs are carefully described in the B.S. and in addition, Loch Bhealaich, which runs into Loch Coire, and an unimportant Loch Syre which joins the River Naver on its right bank about four miles below Loch Naver.

To begin then with the five measured lochs in the basin.

Loch Bhealaich is less than 2 ft. above Loch Coire and is connected with it by a stream 200 yards long:-

> To the north of the two lochs Ben Klibreck slopes gently up to over 3000 ft. while the ground to the south is not so high, but much steeper; so steep, that around Loch Bhealaich (which lies in a very fine corrie) even at noon on the day of the survey the sun could not be seen, except by going over to the north-west shore. (B.S.).

Loch Bhealaich exceeds 1½ miles in length and drains an area of six square miles. The maximum depth of 80 ft. was observed towards the north-east end of the loch and the mean depth was over 30 ft. The elevation was found to be 527 ft.

The main body of the loch is quite simple in formation. At the north-east end there is a small expansion of the loch having a maximum depth of 14 ft., separated from the main body by a constriction in which the depth is 9 ft. The 25-foot area is over a mile, and the 50-foot over half a mile in length, the deeper water being contained in the north-eastern half of the loch, the deepest sounding of 80 ft. having been taken about quarter of a mile from the north-eastern shore.

Less interesting is Loch Coire into which Loch Bhealaich enters, making with it a loch of 4¾ miles in length. The maximum depth of this loch is 151 ft. and was observed comparatively near the south-west end; the mean depth is nearly 59 ft:-

> The loch is quite simple in conformation, with the deeper water lying towards the south-west end, and the fact that in Loch Bhealaich the deeper water also approaches the separating peninsula seems to suggest that the two lochs may at one time have been continuous. (B.S.).

Loch Meadie is very regular in outline, the northern portion being narrow, while the southern portion widens out considerably; there is a very narrow and shallow constriction near the middle which practically cuts the loch into two portions. It is 3⅓ miles in length, and has a maximum breadth near the southern end of over a mile, the breadth of the

entire loch being a quarter of a mile. The maximum
depth of 63 ft. was observed in the wide southern
portion of the loch.

The loch is very irregular in conformation, with
many small islands in the southern half. A sounding
of 44 ft. was taken about 200 yards from the
southern shore, and there is a small area about one-
third of a mile in length exceeding 50 ft. in depth.
On the whole, Loch Meadie is rather shallow, 70 per
cent of the lake floor being covered by less than 25
ft. of water. The height of the loch is 488 ft.

Each of these three lochs holds salmon and sea trout
and also ferox, ordinary trout and char, as also does
Loch Naver.

I may now describe Loch Naver itself, which is the
lowest loch in the basin. It lies about five miles to
the south-east of the last loch, with Ben Klibreck to
the south rising gently from the shore. It is broadly
sinuous in outline, the general trend being east-
north-east and west-south-west, while the upper
portion lies east and west. It exceeds six miles in
length. The maximum depth of 108 ft. was observed
in the wider part of the loch towards the west end.
The height above sea level is 248 ft. and the mean
depth is 39 ft.

The floor of the loch is rather irregular. The 25-foot
contour line is discontinuous opposite the entrance

of the Allt Gruama Beag, where the deepest
sounding was 24 ft. The 50-foot contour is
continuous, enclosing an area nearly four miles in
length, distant from the east end about 1¾ miles and
approaching to within a third of a mile from the
west end. The main 75-foot area is three miles in
length and approaches to within three-quarters of a
mile from the west end. There are two very small
100-foot areas based upon isolated soundings of 100
and 108 ft., the former opposite Gruama Mor, the
latter further up the loch opposite Reidhacheisteil:-

> On the northern shore of the loch, Reidhacheisteil
> and Gruamamore, and on the southern shore,
> Ruighnasealhaig, are the ruins of considerable
> villages destroyed at the beginning of the last
> century, when the crofters were turned out. (B.S.).

Loch Syre is the fifth and last measured loch in the
basin. It is at a height of 413 ft. and lies on the high
ground between Strathnaver and Loch Loyal, and
enters the River Naver by the Langdale Burn four
miles below the River Mallart on the opposite bank.
It is an irregular shallow loch with several islands
and the eastern part is full of stones. The maximum
depth is 12 ft. and the mean 5½ ft. It has small trout,
but red-fleshed.

There are other lochs in the Naver basin which
should be named at least and some of these must
hold char.

(1) Three small lochs on the Allt a' Ghlas-locha, a tributary of the River Mudale - Loch Ben Harrald, Loch an Glas-loch and Loch Eileanach.

(2) A small loch, Loch Gaineamhach (1242 ft.) which runs into the head of Loch Bhealaich.

(3) Three lochs that are on a small burn that runs into the River Mallart, Loch Coir nam Feuran, Loch Ruigh nan Copag, and the highest, Loch Tarbhaidh, with half pound trout or larger.

(4) On a burn running into the north side of Loch Naver with three lochs on it - Loch Molach, Loch Gruama Mor and Loch Eileanach (the highest, stocked with trout about 40 years ago, of considerable size).

(5) Under the head of Ben Clibreck, Loch nan Uan, where rainbow trout were stocked 40 years ago, runs into the south end of Loch Naver. In a neighbouring catchment, Loch na Glas-choille and Loch Bad an Loch, also run into Loch Naver. All three lochs are reputed locally to have char.

(6) Rhifail Loch, on a burn running into the River Naver (right bank).

(7) On Skelpick Burn (right bank of River Naver) are two lochs - near its head, Loch Rifa-gil, and

lower down, Loch Mor na Caorach (trout averaging
2 lbs.).

(8) A small loch, Lochan Druim an Duin, with good
trout, in which one of 11½ lbs. was got. The loch is
high, yet sandy; the sand apparently being blown up
from the mouth of the River Naver.

Borgie

We have next to attack the basin of the River Borgie
which enters the sea more than a mile to the west of
the River Naver. Here there is a well known chain of
lochs - Loch Slaim (360 ft. above sea level), Loch
Craggie (369 ft.), Loch Loyal (369 ft.) and Loch
Coulside (387 ft.) - in order of ascent from the sea.
All of these lochs except the first were surveyed by
the B.S. It, we may dismiss as being shallow and
mainly good for salmon. The last also, need not be
considered, its maximum depth being only 14 ft.
There remain then Lochs Craggie and Loyal which
are practically connected: the maximum depth of the
former is 85 ft., that of the latter 217 ft.

Loch Loyal is well known to contain char (no doubt
there are also char in Loch Craggie). Loch Loyal sits
at a height of 370 ft. and is 4½ miles in length with a
maximum depth of 217 ft., which was observed near
the foot of the loch, little more than half a mile from
the northern shore and a mean depth of 65¼ ft.:-

To the west of the loch rises Ben Loyal, the
highest point exceeding 2500 ft. in height. To the
east of the northern portion of the loch rises Beinn
Stumanadh to a height of 1728 ft., along the base
of which the shore is thickly wooded. In outline,
the loch resembles somewhat a wellington boot,
with the toe pointing in a westerly direction, while
the body of the loch trends almost north and south.
(B.S.).

Loch Loyal contains two deep basins, the larger and
deeper in the northern portion of the loch and the
smaller and shallower towards the head of the loch,
separated by a shoaling of the bottom about 2½
miles from the foot of the loch where there is a
slight constriction in the outline. The 50-foot
contour line is continuous, and encloses an area of
about four miles in length extending from quite
close to the northern end to within half a mile of the
south-western end. There are two 100-foot basins;
the smaller one approaches to within less than a mile
of the head of the loch and is three-quarters of a
mile in length, the maximum depth therein being
137 ft., about a mile and a half from the south-west
end. The larger basin is over two miles in length and
approaches to within 250 yards from the northern
end, enclosing the deepest part of the loch. The 150-
foot area is about 1¼ miles long, and distant about a
quarter mile from the northern end The 200-foot
area is nearly three-quarters of a mile in length,
distant less than half a mile from the northern end.

We find an account of the fish in the loch in the
O.S.A.:-

> Loch Laoghal [Loyal] and the Torrisdale [the
> name of a village and bay at the foot of the
> Borgie] produce fine trouts, eels and salmon. The
> char (tardeargan) appear in shoals in October, and
> are caught with net only.

In default of old records I may quote a
comparatively recent (1878) impression of fishing
on Loch Loyal from the fourth edition of J.
Colquhoun's *The Moor and the Loch*:-

> The lochs were at their lowest, consequently very
> difficult to troll. There is only one state that I
> consider worse - viz. when too large. You may
> then proceed with fly - trout and very small
> feroxes, but until the lochs fall very considerably,
> the only ones worth hooking seem glued to the
> bottom ... The biggest ferox captured by us on
> Loch Layghoul was secured by the worst-baited
> we ever insulted them with. Having used up all the
> best live bait; there only remained one or two
> small things quite unfit for trolling. I had landed at
> the Shepherd's Brook for a fresh supply, while my
> son put on a couple of these tiny trout with
> scarcely an attempt to make them spin. Very soon
> this large ferox dashed at the clumsy lure, and
> after a wicked struggle, was persuaded into the
> landing net and proved a large-headed, ill-shaped
> monster of 14 lb. The next fish we landed was a
> perfect contrast - viz. a beautiful specimen of 11½

lb. and in the finest order, on which account my
son had him preserved for our collection. On the
other days when we trolled the loch our bags
generally contained from two to three feroxes of
from 6 to 8 lb. ...

We trolled Loch Naver, a five pound trout, three
two-pounders - in size, shape and colour exactly
alike - and several near a pound, were the day's
produce. We saw nothing of the salmo-ferox in
Loch Naver, but this was the only day we tried it.

The depth of Loch Loyal is shown clearly by the
following list of temperatures at different depths,
which were taken by the B.S. on September 29,
1902, at 12 o'clock.

Depth	°F.
Surface	53.0°
25 ft.	52.5°
50 ft.	52.5°
70 ft.	52.4°
100 ft.	52.4°
125 ft.	52.9°
135 ft.	47.8°
145 ft.	46.7°
150 ft.	46.5°
195 ft.	46.1°

Loch Coulside trends east-north-east and west-
south-west and is very nearly a mile in length. The

maximum depth is 14 ft. and the mean 7½ ft. The maximum breadth being about 250 yards. The upper portion of the loch is being silted up and the lower is full of stones.

Loch Craggie (or Creag) is over 1½ miles in length and its maximum depth is 84 ft., its mean depth being 33 ft. Like Loch Loyal, it contains two deep basins which are separated by shallower water at the constriction in the loch towards its northern end. The deeper basin occupies the wide southern portion of the loch, towards the peninsula separating this loch from Loch Loyal, in which also the deeper water approaches the dividing peninsula, suggesting that at one time the two lochs might have formed a continuous sheet of water.

The principal 50-foot area is about three-quarters of a mile in length, located less than a quarter of a mile from the southern end of the loch. The maximum depth of the loch (84 ft.) occurs about three-quarters of a mile from both ends but towards the western shore. Towards the northern end of the loch lies the second 50-foot area, based on soundings of 50 and 51 ft., and of small extent, the greatest depth recorded on the ridge separating the two deep basins being 26 ft. close by the eastern shore.

The following small lochs in the basin of the River Borgie should be noted.

(1) Loch Slaim, mentioned above, should be separately enumerated as the first in the chain of lochs on the River Borgie. All have salmon, sea trout, ferox, ordinary trout and (probably) char. But Loch Slaim is especially important for salmon, as it is the first entered by them and probably most are caught in it.

(2) Loch nam Breac Buidge, on a tributary entering the River Borgie on its left bank, about four miles from Tongue, has trout averaging 1 lb. There are also small lochs to the north with trout between 1¾ and 4 lb.

(3) Halfway to Loch nam Breac Buidge from Tongue is a small loch with trout of 2-4 lb.

(4) na Caol-lochan and other lochs, east of the River Borgie, on the Allt Borgidh Beag, have some good trout.

(5) Loch na Moine was drained about 1880.

(6) Loch na Beiste is on a small burn entering the River Borgie between Loch Coulside and Loch Loyal; it is said to have large trout.

Moving west from the River Borgie we have a group of small lochs near the Free Church manse north of Tongue, on the small Allt an Dearg basin.

Kinloch

We may next proceed to the basin of the Kinloch
River.

Lochan Hakel, south of Tongue, with small trout, is
in the small basin of the Rhian Burn, which falls
into the sea at Tongue. There are many lochs
between it and the River Borgie.

Two lochs in the Kinloch Basin have been sounded
by the Bathymetric Survey.

One is a considerable loch, about a mile and a half
long, Loch an Dherue, so called presumably because
it was on the margin of the old Dirriemore, the 'great
forest' of the local aristocracy. The loch lies between
Ben Loyal and Ben Hope at a height of 268 ft. and is
very deep for its size. The general trend of the loch
is north-north-east and south-south-west, the main
body trending almost north and south and throwing
out an arm to the north-east. The loch is over 1½
miles in length, the maximum depth is 157 ft. which
was observed near the centre of the main body of the
loch; the mean depth is 66 ft.

Loch an Dherue has two basins. Firstly, a larger
deep basin in the main body of the loch and
secondly, a smaller shallower basin in the north-east
arm of the loch separated by a low ridge on which
the deepest sounding was 49 ft. The maximum depth

observed in the smaller basin was 59 ft. The 25-foot contour is continuous from end to end of the loch, coinciding approximately with the outline of the loch but approaching close to the eastern shore.

The 50-foot area, in two portions, has the main basin approaching close to the southern end of the loch and exceeds a mile in length. Along the eastern shore, the slope of the bottom in places is very steep. Off Creag an Dherue one sounding of 65 ft. was taken just 100 ft. off shore, and another sounding of 65 ft. about 60 ft. off shore, thus the angle of slope exceeds 45°. Loch an Dherue has salmon and trout, there is, however, no record of char in it, nor in the group of small lochs which run into it, about thirteen in number.

Temperatures on October 1, 1902, were as follows:

Depth	°F.
50 ft.	53.0°
100 ft.	52.5°
125 ft.	48.6°
145 ft.	48.4°

Loch Haluim lies to the south-west flank of Ben Loyal, little more than a mile from Loch Coulside and runs into Loch an Dherue. It is 696 ft. in height above sea level, and is 'most irregular in outline and in conformation, with one comparatively large

island and weeds obstructing many of the bays.' The
maximum depth of 30 ft. was observed in the
extreme western portion of the loch, the mean depth
being about 8 ft. There are two 10-foot basins, the
principal one extending from the extreme end of the
loch to beyond the island. It has good trout, running
up to nearly half a pound.

There are other lochs near Loch Haluim, of which
Loch an Ealachan is the largest (nearly three-
quarters of a mile long), but Loch an Aon-bhric
(One Trout Loch) is alone worth noting. It holds
good trout, said to run up to seven pounds.

After leaving the mouth of the Kinloch River, we
have to pass through no fewer than eight (or so)
small river basins on the west side of the Kyle of
Tongue, and going round Whiten Head. Most of
these basins contain a small loch, but the lochs that
one may name are Loch a' Mhulinn near Melness,
over half a mile long, and Loch Maovally on the
Allt an t-Srathain, which takes us into the
considerable Hope basin.

Hope

Only one loch has been measured here, Loch Hope,
which is close to the east shore of Loch Eriboll at an
elevation of only 12½ ft. above sea level. Thus a
slight subsidence would convert it into an arm of
Loch Eriboll, and at the upper end of the loch, three

terraces are to be seen, and traces of perhaps a
fourth.

Ben Hope rises very steeply to a height of over 3000
ft. to the south-east of the head of the loch, and
some parts of the shore are well wooded. On the
date of the survey (September 30, 1902) a reputed
old castle was just showing a few inches above the
surface about a mile from the foot of the loch.

The trend of the loch is almost north and south and
the length exceeds six miles. The maximum depth of
187 ft. was observed about midway between the two
ends of the loch, and the mean depth is about 61½
ft.

Proceeding from the lower (northern) end of the
loch for a quarter of a mile, one meets with a small
25-foot area, which lies towards the eastern dshore.
Off the opposite shore here, there were many
boulders in the water. Proceeding southwards, the
water rapidly deepens until it attains a depth of 104
ft. opposite the entrance of the Allt an Ruighein,
about 1½ miles from the foot of the loch. Thence,
for a distance of about a quarter of a mile, the
bottom rises again until the depth in the centre is 44
ft. This shoal coincides with a narrowing in the loch,
whence to the south it broadens out again and the
water deepens rapidly to its maximum. The off-
shore slope is in some places rather steep, for
instance along the eastern shore, off the entrance to

the Allt a' Mhuilinn, a sounding of 53 ft. was taken
about 60 ft. from the shore. Off the entrance to the
Allt Braesgill a sounding of 28 ft. was only 30 ft.
from the shore.

Loch Hope was the only loch in the basin that was
measured but there are many others which may at
least be indicated, and there is one very small loch
which is well known, as it contains char.

In the first place, right at the head of the Strathmore
River, which runs into Loch Hope, near Ben Hee
and in the pass called Bealaich nam Meirleach,
which divided off the western section of the
mediaeval 'great forest' (Dirriemore) there is a pair
of connected lochs which together measure more
than a mile and a half – Loch an t-Seilg and Loch an
Aslaird.Then, there are two lochs in the corries of
the same section of the old forest (now the Reay
Forest), Coire Loch and Loch na Seilge, near Saval
More, each of which is half a mile long; both run
into tributaries of the Strathmore River.

The small lochs on the east side of Loch Hope are
most interesting and especially the smallest of them,
about which we have further information. It has an
extremely small drainage area and no visible outlet.
The loch seems to have been first mentioned by F.
Day who tells us in his *British Salmonidae* (1887)
that 'there is a char loch about twenty five miles
from Durness, on Ben Hope, where they are taken

up to 1½ lb. weight.' Mr Tate Regan gives the following authoritative account of the char in the loch:-

> The Large-mouthed Char (*Salvelinus maxillaris*), of a small isolated loch [Lochan Coiŕ á Ghalaich] under Ben Hope in Sutherland-shire, is a very distinct species, to which I have given the name *S. maxillaris* on account of the notable length of the maxillary which extends back far beyond the eye in adult males; the head is longer and the interorbital region narrower than in the Windermere char.
>
> The colour of the back and sides, with the dorsal and caudal fins is dark plumbeous; belly brilliant orange; small orange spots on the sides, mostly below the lateral line; pectorals greenish with a red margin; pelvics and anal, reddish with a white anterior edge; caudal with an orange margin.
>
> Eleven examples sent to me measure 8 to 11 inches in total length.

One or two of the larger lochs in the same group are said also to have char.

The small River Polla which runs through Strath Beag into the head of Loch Eriboll. It has sea trout and a few salmon and rises in the considerable Loch Staonsaid, more than half a mile long, at a height of

585 ft., with two smaller lochs running into it. It is
said by the S.G. to contain trout.

At this point in our progress round the north coast
we have to note a regrettable gap in the Bathymetric
Survey. No river basin is given all the way
westward to Cape Wrath and southward from there
to Laxford. This means missing two rather well-
known fishing centres, namely those which were
operated from hotels at Rhiconich and Durness.

Durness

Entering the village of Durness, we pass the well
known Smoo Cave, through which a burn from
Loch Meadie (three-quarters of a mile long)
descends. (See Tate Regan on the trout in the Smoo
River). Within a mile we cross two small streams
which come from three lochs of a very interesting
character thus described by Pont or some other early
geographer (W. Macfarlane's *Geographical
Collections*):-

> The loches of the parish of Durness besides these
> already named and from which there run several
> burns and rivulets to the sea are viz the Loch of
> Slaness ¼ mile in circuit. It stands by the farm of
> that name above mentioned; it hath plenty of eels.
> A little island in it where maws lay eggs. A little
> stripe runs from it to Lochborely [Borralie] which
> bears S.E. of the former a quarter of a mile and is
> half a mile in circuit stands by the farms of Borley

and Claiseneath, it hath plenty of red belly'd trout, an island also where fowls lay their eggs, a burn runs underground out of this loch for a quarter of a mile and falls into another little loch a quarter of a mile in circuit calld Loch Crospuill [Croispol], near to the church and Lord Reays Mannour House, and out of which there runs a burn into the sea. The 3rd loch is called Loch Calladail [Caladail], a large mile in circuit, and is about a miles distance from the Mannour Hose last mentioned bearing S.E. It has plenty of excellent trouts and a burn runs out of it to the sea at Sangomoar.

This is a careful description of the topography of the four Durness lochs about three hundred years ago, and we have to give an account of their state at the present time. The writer begins with the highest of the chain of three lochs which enter the sea at Balnakeil Bay which was part of the Mannour of Lord Reay, and which seems also to have been a kind of seminary for the Church. He assigns to the loch, which is now called Loch Lanlish, only eels, and 'maws' [gulls]. The water from it to the second loch, Loch Borralie now goes more or less underground, and it contains very fine trout up to 12 lb., probably larger trout than any other loch in Scotland of the same size, the feeding (sticklebacks, etc.), which may have been imported, being very good.

This third loch, Loch Croispol, which at one time
seems to have contained only char, now contains
also good trout, up to several pounds. Borralie runs
underground into Loch Croispol, the lowest loch of
the chain. This loch is famous for its trout and Tate
Regan quotes a description of them: 'in Loch
Crassapuil, which has a sandy bottom, they grow to
a large size, and are silvery with a white belly, and
the back and sides greenish with small dark spots'.
P.D. Malloch has photographs of specimens of them
from 2 to 5 lb. (*Life History of the Salmon, Sea
Trout and other Freshwater Fish*).

The fourth Durness loch, Loch Calladail, enters the
sea by a small burn, not at Balnakeil as the three
others do, but as Pont says at 'Sangomoar', between
the Smoo Cave and Balnakeil. The loch has had a
curious history. At some time, before the O.S. map
was published (1880) it was drained. Since then it
has been filled again with water, and serves as a
water supply for the village of Durness, and again
has 'plenty of excellent trouts', several of which
have been caught at a weight of 9 lb.

All the four Durness lochs are prolific to a high
degree, owing no doubt to the limestone of the
region.

I now go on to the River Dionard, which rises in
Loch Dionard, at the foot of Foinaven, and
eventually enters the Kyle of Durness. Loch

Dionard is in turn fed by five small lochs to the
south and south-east, including an Dubh-loch and,
running into it, Loch Ulbha. Stoddart tells us that
'when passing the Dionard, in 1850, I took a few
throws below the bridge at Achintoul and secured a
fine salmon and several sea-trout', but 'by far the
best pools are high up, a mile or two below where it
issues from the loch.' He seems to have been
interested to hear that 'there is a cave at the side of
Loch Dionard, in which the Lords of Reay used to
pass the nights when shooting in the adjoining
forest, or fishing in the lake and river.'

We now, in our progress round the coast of
Scotland, cross the River Dionard and enter the
region called the Parph. This was the waste which
stretched westward from the River Dionard to the
Atlantic, and to Cape Wrath in the north-west. It
seems to have been an outlying part of the 'great
forest' of Reay which was used mainly as a trap for
deer, a purpose for which the angular configuration
of the land made it specially adapted. The main seat
of the Barony was at Tongue and while the central
part of the forest of Reay consisted chiefly of Saval
More, Saval Beag, Arkle, Ben Stack and Foinne-
Beinne (where, it was said, the deer were so prolific
and so flourishing that they had three tails). It
included also Ben Loyal and Ben Hope and, as we
have seen, the Parph.

The coast, from Durness to Cape Wrath, and from the latter to Loch Inchard and Rhiconich, traverses or contains some seven or eight small river basins which contain several lochs of considerable size. For instance, between Durness and Cape Wrath there are Loch Airigh na Beinne and Loch Inshore, neither, however, of any special importance.

Then when we pass the turning point (Cape Wrath) and proceed southward, we find that the B.S. gives no loch that runs into the sea between Cape Wrath and Rhiconich. We first come to the Keisgaig River with Loch Keisgaig, said to have some good trout. Next, on a considerable stream that holds salmon and sea trout, the River Shinary and Sandwood Loch, a mile and a quarter long, which is situated almost on the beach. Sea trout have ready access to the loch and for some considerable distance upstream. There are more than twelve smaller lochs which are connected with Sandwood Loch of which nothing is recorded. Only one or two of them are above 700 ft. The surrounding hills are in what is called the Parph, part of the mediaeval forest mentioned above.

V

LAXFORD, SCOURIE, BADCALL AND DUARTMORE

Before we arrive at Rhiconich, we pass three or four small river basins, each of which contains several lochs. The two largest of these lochs are nearly a mile in length - Loch Innis na Ba Buidhe at Kinlochbervie and Loch na Claise Carnaich near Rhiconich; neither of these is over 500 ft. in altitude.

We have now reached Loch Inchard and the region round Rhiconich. Here there are very many lochs, but they are mostly small and unimportant, so we must be content with a selection. The only notable stream, the Rhiconich River, enters the sea at the head of Loch Inchard near Rhiconich. It holds salmon and trout and has two lochs on it, first Loch a' Garbh-bhaid Beag and then Loch a' Garbh-bhaid Mor above. The former is over half a mile in length and not deep, while the latter is a mile long and presumably fairly deep; A. Grimble says it has char as well as sea trout.

Halfway along Loch Garbh-bhaid Beag on the north side there enters a large burn, the Garbh Allt, which descends from a chain of five lochs, called no 1, no 2, no 3, no 4, and no 5, in the nomenclature of the inn at Rhiconich. They are at heights running from 400 ft. to 700 ft., and there is a small tarn still higher (above 1250 ft.) which may count as a 6th in the sequence of lochs; it is very near the summit of Arkle (2582 ft.). All the five lochs contain trout at least, and no 2, which is more than a mile long, is said to have ferox up to 9 lb.

The other stream near Rhiconich that is of interest is further west and falls into the sea near Loch Laxford. There are two lochs on it which have sea trout and it rises in Loch na h-Ula (Loch na Thull), which holds ferox up to 7 pounds or so, and char. The burn crosses the road about a mile west of Rhiconich and falls into the sea at the head of Loch a' Chad-Fi, a small fjord which is a branch of Loch Laxford (= lax-fjord). The burn passes through two sea trout lochs on its way to the sea.

The char loch, Loch na h-Ula, is not given in the B.S., though it is more than a mile long. It is not high, but there are small lochs running into it, which might be worth investigating.

By following the road from Rhiconich to the head of Laxford, we are omitting the peninsula that lies to the west of Loch Inchard, which holds innumerable lochs. The largest of these is Loch Crocach, which is more than a mile long but is not in the B.S. It is on one of the lochs on this remote promontory that St John is said to have made observations on the Osprey.

At this point, the gap in the B.S. comes to an end, and we have the guidance of its scientific measurements of the lochs on most of the rest of the coast.

Laxford

We have now reached the Laxford basin. The geography of it is fairly simple.

The source of the River Laxford is Loch More, the largest loch in the Laxford basin, which is four miles long, and is very deep, a maximum of 316 ft., and a mean depth of 126 ft. The loch is described in the B.S. as 'quite simple in conformation' while 'it partakes of a flat-bottomed character, as evidenced by the comparatively large area of the lake floor covered by more than 200 ft. of water.' It lies at a height of 127 ft. above sea level.

The record of temperature is instructive, for in the deepest part of Loch More on September 6, 1902, the position of the 'discontinuity layer' was evident.

Depth	°F.
Surface	54.4°
100 ft.	51.2°
103 ft.	50.6°
106 ft.	47.6°
290 ft.	45.7°

The second loch examined is Loch nan Eulachan, at the same level as Loch More and immediately below it. The loch is over 50 ft. deep, and is to be regarded rather as a part of Loch More rather than as a separate loch.

Then, after a short run of less than a mile, the Laxford River enters Loch Stack, the form of which is described as 'having a fanciful resemblance to the capital letter H, with one arm longer than the other; in fact it may be looked on as two lochs joined by a shallow neck.' 'A line along the axis of maximum depth (108 ft.) and along the axis of the other arm would be 3¼ miles in length.'

The mean depth of the loch is 36 ft., and in the shallow joining the two arms of the loch the depth is 16 ft. More than 70 per cent of the lake floor is covered by less than 50 ft. of water, and 40 per cent by less than 25 ft.

Each of two burns that enter Loch Stack has on it a fairly big loch, the larger, Loch na Mucnaich, being nearly a mile long, and the smaller, Loch an Nighe Leathaid, on the slope of Arkle, over half a mile.

The statement in the B.S. that 'Loch More contains splendid trout, while Loch Stack contains also sea-trout, Salmo ferox, salmon and char' perhaps sufficiently describes the fishing in these lochs, except that the ferox and the char are probably

chiefly in Loch More and that there are sea-trout in
it as well as in Loch Stack. Stoddart tells us that
ferox up to 25 lb. have been got in Loch More, and
he says that on Loch Stack in 1950 'in the course of
a few hours, and without a landing net, I captured
thirty-one sea-trout, the largest upwards of five
pounds weight; several yellow trout.'

After a further course of four miles or so in a north-
west direction, the River Laxford enters the sea at
the head of Loch Laxford.

Mr Tate Regan classes the char of Loch Stack, in
one respect at least, with those of Scourie and the
tarn on Ben Hope, as having a comparatively small
number of vertebrae. Nothing seems to have been
recorded about their size and appearance.

Scourie

We are now approaching the district of Scourie. The
village of Scourie is about six miles west of Laxford
Bridge at the head of the fjord. The geography of the
district is rather complex and it is most convenient
to distinguish four small basins, each with its
corresponding short river.

(1) The River Sealbhag rises in Gorm Loch at a
height of nearly 500 ft., passes through Loch Bad an
t-Seabhaig (under 100 ft.) - a sea trout loch - and
then enters the sea at the head of Laxford Loch.

Loch Bad an t-Seabhaig has been surveyed and has a maximum depth of 56 ft. There are several other, smaller, lochs in the basin.

(2) The B.S. includes in the Laxford basin, two lochs near Scourie which enter the sea on the south side of Laxford by different burns. One of these is Loch na Claise Fearna which has a few sea trout and has been surveyed by the B.S. It has a maximum depth of 38 ft. In addition there are a few smaller lochs in the basin, the highest of which, Loch a' Mhuirt, is under 400 ft.

(3) The third basin contains Loch nam Breac, an interesting loch and the largest in the Scourie district. It is a rather irregular star-shaped loch which is described as 1⅛ miles in length. The bottom sinks in two places below the 50-foot level: (a) in the most southerly expansion of the loch, where there is a basin a quarter of a mile in length, enclosing the maximum depth of the loch (71 ft.); (b) to the south-west of the largest island, where there is a smaller basin having a maximum depth of 66 ft. It lies at about 250 ft. above the sea.

The temperatures on September 9, 1902 are interesting. 'This series shows a range of 6.1°, the position of the sprungschicht [a thermocline] being well-marked.'

Depth	°F.
Surface	57.1°
45 ft.	56.0°
48 ft.	52.0°
60 ft.	51.0°

The loch contains, on good authority, char, ordinary trout and possibly ferox. There is a small loch on the burn between it and the sea.

(4) The fourth basin drains into Scourie Bay and contains two lochs which have been investigated by the B.S. They are in what it describes as 'the Scourie basin', and raise a difficulty.

The upper of them, Loch Laicheard, which runs into the lower, Loch a' Bhadaidh Daraich, is nearly a mile long and on the whole, shallow, having a maximum depth of 42 ft. and a mean of 15 ft. It is at a height of nearly 400 ft. and runs down to Loch a' Bhadaidh Daraich, which is only 48 ft. above the sea, being connected with it by the short Allt a' Mhuilinn, running into Scourie Bay. Loch a' Bhadaid Daraich is nearly a mile long and is deep, 121 ft., the mean depth being considerable - 55 ft. It forms a simple flat-bottomed basin, and there are in places steep off-shore slopes, as, for instance, off the northern shore towards the west end, where a sounding of 45 ft. was taken about 40 ft. from shore.

The latter loch has small ordinary trout and a few sea-trout.

We are now in a position to consider the rather obscure question of char at Scourie. The first reference to Scourie in Mr Tate Regan's *The Freshwater Fishes of the British Isles* is on page 75 where he is contrasting 'char lakes situated at an altitude of more than 1000 ft.', as on Ben Hope with 'Loch Scourie in Sutherland' and others, which are less than 50 ft. above the sea. In a later passage he says, in an enumeration of species:

> Malloch's Char (*Salvelinus mallochi*) is another Sutherlandshire form which occurs in Loch Scourie where four examples, 8 to 12 inches long, were captured by Mr P.D. Malloch, the well-known naturalist of Perth, in whose honour I named the species.

In point of fact, no loch in Sutherland seems to be called Loch Scourie. But Loch a' Bhadaidh Daraich, which is at Scourie, satisfies Mr Regan's description of the loch in question, as being less than 50 ft. above the sea. Loch nam Breac on the other hand, which also has char, is at a considerable height. Further, all that Malloch says in his *Life History of the Salmon, Sea Trout and other Freshwater Fish* (published in 1910, the year before Mr Regan's work) is that 'at Scourie, on the west coast of Sutherland, some of the lochs contain large char.'

Clearly, we may guess, that Malloch caught char in both lochs.

Mr Regan's description of Malloch's Char is as follows.

> This is a rather short-headed, blunt-snouted, and small-mouthed char, with small scales (188 to 200 in a longitudinal series). The lower jaw is obtusely pointed anteriorly and slightly shorter than the upper even in adult males, the paired fins are short. The back and sides are slate-coloured, covered with numerous pale spots, the belly whitish, tinged with orange.

The Scourie area and surrounds is the most interesting fishing district in Scotland, if you want to catch large trout with fly. There are at least 56 lochs within moderate distance of the village, in which, with a reasonable amount of skill, you have a good chance of killing a trout of over 2 lb. which cannot, I think, be said of any other region in Scotland. In many of them you have a chance of getting a 6 lb. trout or even a larger fish.

The Scourie lochs are inseparably associated in the minds of many Scottish fishers with Cuthbert W. Finch, who visited them every year of the second and third decades of the present [20th] century, during the months of May, June and July. He had evolved a technique which was appropriate to the

district and applied it with consummate skill. His
main principle was that you should see a large trout
before you try to catch it. Thus, in one way, the
method resembled dry fly fishing in a river, but here
you were dealing with small lochs, not with pools in
a river. Thus Finch and his *Fidus Achates*, Donald
Macrae, would approach the loch that was their first
objective for the day and take up their positions at
places which were at a considerable distance from
each other, say on different sides of the loch, or at
different ends. Then each would report to the other
by means of agreed signs (such as the calls of birds)
whether any large trout were moving in the loch,
and if so, where. If nothing was moving, the
confederates would go on to another loch that was
near and had large trout, and attack it in the same
way. If nothing was rising at all, anywhere, they
would fish one or two of the lochs in the usual way -
i.e. 'chuck and chance it'. But as a rule they would
find some fish somewhere, and could fill the whole
day with a round of lochs. For instance, they might
begin with Upper Calva and then go in succession to
Lower Calva and Strathan, ending up perhaps with
Black Rock, which is within an easy distance of the
village. Both were keen observers of nature. They
took with them simply two fishing rods, though it
was only rarely that they were both fishing at the
same time.

The method, as we should expect, was extremely
effective, indeed in a way, if one may criticise, too

effective, and resulted in the capture of many large trout. The rod used was short, but very powerful and at each cast, two yards or so of line were drawn in by hand and shot out through the rings in casting. The line went out very straight, but the cast was perhaps neither very light nor very neat, but rather resembled throwing or casting a light spinning bait, and drawing it out very carefully by hand.

Finch usually fished alone (with Donald) but sometimes, if Donald had to give a day to his croft, he would ask a friend to accompany him. This friend was carefully selected by Finch.

Badcall

The next basin examined by the B.S. is the Badcall basin. Only one loch was investigated

Loch na h-Airigh-Sleibhe, commonly known to Scourie fishers as the Fairy Loch, lies little more than half a mile to the south-east of Loch a' Bhadaidh Daraich. It stands fairly high, nearly 300 ft. and runs down into Loch Bad nam Malt, which has a few salmon and sea trout, and enters the sea at Badcall Bay, two or three miles south of Scourie.

Loch na h-Airigh-Sleibhe is two-thirds of a mile in length and has a maximum depth of 113 ft., and a mean depth of 44 ft. The shallower contours coincide approximately with the outline of the loch and in some places deep water approaches close to the shore, as, for instance off the northern shore, in the vicinity of the deepest part, where a sounding of 43 ft. was taken about 40 ft. from the shore.

The series of temperatures on September 10, 1902, is interesting.

Depth	°F.
Surface	56.4°
40 ft.	56.4°
45 ft.	53.6°
50 ft.	51.1°

For fishing, a more important tributary of the Badcall River enters Loch Bad nam Malt from the east, rather than from the north-east. This tributary has a chain of small lochs on it. The majority of these hold large trout and the more important of them ought at least to be mentioned. The source of this tributary is firstly Loch Eileanach, or rather a very small loch that runs into it. Then, coming downstream, there is Lochain Doimhain (with a very small loch appended to it and containing big trout). These lochs are especially interesting as they are said, on good authority, to contain char, but they have not been sounded. Then, still further down, there is Loch nan Uidh, and running into it Lochain na Creige Duibhe (Black Rock).

Lastly, further south, on the Allt an t-Stathain, there is a considerable loch, Loch Crocach, which was not surveyed by the B.S. It is over three-quarters of a mile long and lies at the foot of Ben Auskaird.

Duartmore

We now return to the safe guidance of the B.S. which describes two lochs on the Duartmore River, which enters the sea a mile south of Calva Bay.

The lower loch, Loch Duartmore, is small - not much over a quarter of a mile long - and shallow, with a maximum depth of only 22 ft. Above it, Loch Eucail, which is 'overgrown with weeds and apparently shallow'. Then there comes Loch Allt na h-Airbhe, which is nearly two-thirds of a mile long and fairly deep, 60 ft. maximum with a mean depth of 30 ft. These lochs have sea trout and salmon.

The Duartmore basin contains three good trout lochs, which I will append to the three that I have named above. They are: (1) Lower Loch Calva which has very good trout up to 6 lbs. or so. (2) Loch a' Mhinidh, with a smaller loch running into it. (3) Little Strathan Loch.

After Duartmore, there is a gap in the B.S., and no basin is given until we reach Loch Roe, two or three miles north of Lochinver. But the whole district near Kylesku and round the Point of Stoer is rather

interesting and a few lochs must be named. Kylesku
is a narrow fjord which has two branches - Loch
Glendhu and Loch Glencoul. At Kylesku we may
distinguish four basins.

(1) The basin of the Maldie Burn, which enters Loch
Glendhu near its foot. The basin has two main
features - a high fall near the foot of the river,
stopping the further ascent of sea trout; and a loch a
mile and a half in length a short distance above the
fall - Loch an Leathaid Bhuain. Close to the head of
the loch is another loch, Loch na Creige Duibhe,
half as long, at three quarters of a mile in length.
The head of the latter is deep and stony but it has
trout. There are several other, smaller, lochs in the
basin, most of them at a height of 1250 ft. or very
nearly.

(2) In the basin at the head of Loch Glendhu there is
a loch, Loch Srath nan Aisinnin, three-quarters of a
mile in length, and several smaller lochs at over
1250 ft.

(3) The basin at the head of Loch Glencoul holds
two considerable lochs. One, Loch nan Caorach, is a
mile long; the other, Loch an Eircill, is half a mile in
length. Again, there are several small lochs at over
1250 ft. and one at over 1750 ft.

(4) Lastly, is the small Unapool Burn running into
the foot of Loch Glencoul (but not including in its

basin the small Loch Unapool itself, which is nearby in a very small basin of its own). Here there is Loch na Gainmhich, half a mile long with three small lochs above it, one of them at a height of 1700 ft. and another nearly as high.

Going westward along the coast towards the Point of Stoer, there is a small river which comes through Gleann Leireag and enters the sea at Loch Nedd. It holds some sea trout and has a considerable loch on it, Loch an Leothaid, which is more than than half a mile long.

Passing near Nedd and two largish lochs, Loch Torr na h-Eigin and Gorm Loch Mor (nearly a mile long), the sources of the Oldany River, which has some sea trout as far up as the fall below Loch Poll, which is 1½ miles long. In the same basin there are Loch Drumbeg, Loch na Loinne and, near the head, Gorm Loch Beag, which has good trout.

Near the top of the hill after crossing the Oldany River, among a maze of lochs, there are two very good lochs: Loch Eileanach and Loch a' Phollain Bheithe. Then, connected with Loch na Bruthaich, which is near the sea, there is a group of small but good lochs, the largest being called locally the Shepherd's Loch.

We now come to the Clashnessie Burn, which, after a waterfall, enters the sea near Clashnessie, coming

from a group or network of lochs. The longest of
these is Loch nan Lub (three quarters of a mile
long), and above it are Loch Poll Dhaidh and Garbh
Loch Mor, the latter being the better for fishing.

The coast between Clashnessie and Lochinver
(round the promontory of Stoer) resembles the
Scourie district, but is less interesting as there seem
to be no char to record. In the Point of Stoer itself
there is Loch Cul Fraoich (nearly three quarters of a
mile long) and Loch na Claise, the former having
been unsuccessfully stocked with Loch Leven trout,
but the latter used to have good trout trout averaging
about a pound.

Turning southward towards Lochinver, we have
Lochan Sgeireach, with very good trout (in 1912),
Loch Leathad a' Bhaile Fhoghair (large trout), and
Loch na h-Uidhe Doimhne (three quarters of a mile
long), with fair trout, and the Schoolmaster's Loch
(stocked with Loch Leven trout).

Further south there are Loch na h-Airigh Bige (over
half a mile long) and Loch an Ordain (also over half
a mile long), and between them a mysterious
nameless loch about half a mile long, which is not
marked in the first (1880) edition of the one-inch
O.S. map, with a small loch running into it. The
latter must have large trout, and a 12 lb. trout is
reported to have been seen in the former.

We have now reached Loch Roe and the end of the gap in the B.S. It will be seen that we have had to interpolate over 30 lochs, all of which hold trout, and practically all of them are at least half a mile long. None of these lochs is reported to have char.

Two lochs in the Roe basin are given in the B.S. The larger, Loch Crocach, is about three miles to the north of Lochinver, and is perhaps the largest loch in the Stoer district, being 1½ miles in length. The Survey says that its 'insulosity   (i.e. the ratio between the area of the islands and the total area of the loch) is probably higher than in any loch visited by the Loch Survey, the lochs most nearly approaching this being Loch Maree and Loch Lomond.' Loch Crocach has a maximum depth of 71 ft. and is at a height of nearly 400 ft.

Loch an Tuirc is on the stream running out of Loch Crocach, The loch is rather shallow (maximum depth 39 ft.) and 'weeds are very abundant in some parts of the loch', which holds some sea trout. The Manse Loch and several other smaller lochs are farther down before we reach Loch Roe.

VI

INVER, KIRKAIG, POLLY AND GARVIE

Inver

Seven lochs have been measured in the Inver basin –
Loch Assynt, Loch Leitir Easaidh, Loch Mhaolach-
coire, Loch Awe, Loch Beannach, Loch Druim
Suardalain and Loch Culag. The first is the primary
loch in the basin, and lies about four miles from the
end of the sea-loch, Loch Inver. It receives the
outflow from Loch Awe, Loch Mhaolach-coire and
Loch Leitir Easaidh and its waters are discharged by
the River Inver, which, after a wild and tortuous
course, falls into Loch Inver.

The ground along the western end of Loch Assynt is
low, but on proceeding westward it becomes higher,
and on the north shore, Quinag attains 2,600 ft.,
Glas Bheinn 2,500 ft. and farther to the south-east
Conival and Ben More (Assynt) reach 3,200 ft., the
last being the highest hill in the county.

On a promontory on the north shore about a mile
from Inchnadamph stand the ruins of Ardvreck
Castle, once a stronghold of the McLeods, the
betrayers of Montrose. The general trend of the loch
is west-north-west and east-south-east, while the
western end bends sharply at Lochassynt Lodge to
the south-west and the eastern end less sharply to

the south-east. It is 6 1/3 miles in length. The
maximum depth is 282 ft. and the mean depth 101
ft. The height is 215 ft. above sea level.

The floor of Loch Assynt is rather irregular; this is
more especially the case in the half lying to the
north of the medial line. The 100-foot contour,
running along the northern shore, is of the most
sinuous character, quite independent of the shore
line, and is in striking contrast to the same contour
running along the southern shore. The 50-foot, 100-
foot and 150-foot basins are continuous areas while
the 50-foot contour basin extends practically from
one end of the loch to the other; the 100-foot basin
stretches from 200 yards from the eastern end to
beyond Rubh'an Alt-toir, where the loch bends
sharply to the south-west, and is five miles in
length; the 150-foot basin extends from about a
quarter of a mile from the east end to Rubh'an Alt-
toir and is 4¾ m in length. The deep channel lies for
the greater part of its course much nearer the
southern than the northern shore; opposite Ardvreck
Castle, however, it crosses over and lies nearer the
northern shore at the end of the loch. The numerous
large bays along both shores are fairly deep.

I may quote a few statements about Loch Assynt
from J. Colquhoun's *The Moor and the Loch* (Vol.
II) in his chapter called 'A raid on Sutherland':-

The salmo-ferox of these three lochs (Loyal,
Craggie and Slaim on the Borgie) are scarcely
equal to those of Loch Assynt at the other end of
the county; while the Loch Shin trout taken by the
troll are decidedly inferior to all those lochs – on
which account I never took the trouble to fish it ...

The district of Assynt is unrivalled for loch
fishing. It is full of lakes, with variety of fishing to
tempt all tastes; tarns teeming with yellow trout to
suit impatient fishermen – and ladies it would
seem – for we actually saw two equipped with
little shining fly rods, ready to fill their creels,
which no doubt they did. A gillaroo loch is only
three miles from the inn, and in Loch Assynt
larger feroxes are more frequently taken than in
any other loch in the county. Although vexed to
have only one day to spare for so fertile an angling
range, we determined to make the most of it, and
throw no chance of a heavy fish away. We
immediately set about catching bait of various
sizes, and despatched a lad for John Sutherland,
the Duke's watcher, asking him to accompany us
next day.

There were no fewer than four fishing-parties at
the inn, each choosing his own sport, and plenty of
it to choose from. All of them had tried the large
loch for the big fish, but the inn people told us that
not one salmo-ferox had been extracted this year.
A little enclosed pond was easily found, to keep
the bait alive during the night, and we soon
whipped out the necessary number to stock it.

We were early astir. Of course the live bait were first visited; and as some of the best had escaped, the fly-rod was again put in commission. My comrade was about to empty our fresh supply into the enclosed pool, when, with eager gesture, he called to me that there was a black rat, with a white throat, walking at the bottom among the trout. I was soon by his side and at the same moment the little black-and-white intruder sneaked out of his bath and took shelter in a hollow of the rock. He was easily dislodged and captured, when the prize turned out to be that rare species, the water shrew. We were told that they were numerous in that quarter, although I never met with one in any other part of Scotland.

Of the survey lochs we naturally consider next Loch Leitir Easaidh or Letteressie which lies immediately to the westward of, and at a slightly higher level than, Loch Assynt, into which it flows by a stream only a few yards in length. The ground around it is low. The waterfall at its western end, from which the loch derives its name, is very fine. Loch Letteressie is considerably over half a mile in length. The maximum depth is 70 ft. and the mean depth 20 ft., while the drainage area is 2⅔ square miles.

The loch is extremely irregular in outline, the main body trending north and south, with an arm running in an easterly direction towards Loch Assynt. The maximum depth observed in this arm was 31 ft., separated by shallower water from the deep basin in the main body of the loch, where there is a small area exceeding 50 ft. in depth towards the western shore.

The fish in the loch are the same as those in Loch Assynt namely salmon and sea trout, ferox and ordinary trout and (presumably) char.

Loch na h-Innse Fraoich, runs into Loch Leitir Easaidh.

According to Sir J.E. Edwards-Moss (*Season in Sutherland*), Loch nan Curran at a height of nearly 2060 ft., on a tributary of the River Traligill, is notable for its abundance of freshwater shrimps.

The next loch we have to consider is Loch Mhaolach-coire. Mulach Corrie or the Gillaroo Loch is situated about two miles to the south-west of Inchnadamph and flows by the Allt na Glaice Moire into the River Traligill, which runs into the head of Loch Assynt near the entrance of the River Loanan. The loch derives one of its names from the supposed resemblance of its fish to the Gillaroo trout of Lough Melvin; in shape the trout are deep and thick and heavy in proportion to their length. The loch trends in a north and south direction. It covers an area of about 20 acres, and its maximum depth is 8 ft., its mean depth being 5 ft., while its height is between 800 and 900 ft.

The region is limestone, and the burn from the loch
is partly underground.

In 1853 Stoddart wrote of the loch as follows:

> Of our numerous Scottish lochs, a great proportion
> of which has been investigated by our naturalists,
> one only is affirmed with any degree of
> positiveness, to contain this species of trout. It is a
> small tarn, situated on a shoulder of Ben More, in
> Sutherlandshire, about three miles from
> Inchnadamph, named Mulach Corry. I visited it in
> 1850 under somewhat unfavourable
> circumstances, during the occurrence of a snowfall
> when the loch was partially frozen, but succeeded,
> both with worm and fly, in securing a few
> specimens, none of which, however, exceeded in
> weight half a pound. In the shape and appearance
> of those fish I was much disappointed, nor did
> their edible qualities approach the reputation given
> them. They were very inferior in all respects to the
> trout of Lochs Awe and Assynt a short way below
> them. Nor did the stomach, when examined, differ
> so essentially in its muscular conformation as to
> induce the conclusion that they were a distinct
> species of trout. The gillaroo in fact of Mulach
> Corry, which is situated upon a limestone rock, I
> have every reason to think is nothing more than a
> variety of the fario or common trout; and that the
> gizzard or indurated portion of the stomach which
> distinguishes it, is entirely the result, not the
> occasion, of its peculiar feeding. This is true at

least, that all freshwater trout engross some measure of testaceous food; and when the opportunity occurs, will greedily devour and feed upon and abundantly thrive upon small shell-fish and horny substances ...

This species of trout, I have been told, was discovered in Loch Garve in Ross-shire by the late Sir Humphry Davy. In his *Salmonia*, however, he states distinctly that 'except in Ireland, he never found a gillaroo trout.' The Loch Garve trout are very like those of Mulach Corry and previous to the partial drainage of the lake had few rivals in point of shape, beauty, and flavour among the finny tribe.

Mr Regan says that:-

In many of the Irish lakes (Loughs Melvin, Mask, Corrib, etc.) the fishermen distinguish with the name 'Gillaroo' a trout which differs somewhat in habits and appearance from the ordinary trout. The most notable things about the Gillaroo are that he is always well spotted with red, hence his name, which is derived from giolla, fellow, and ruadh, red, and that he subsists largely on shellfish and has a remarkably hard and thick stomach.

The fish called 'Gillaroo' in various lakes agree with this, but differ somewhat in other characters, such as form and size, and value as food, according to the locality.

According to Thompson (1856) the Loch Neagh
'Gillaroo' has the upper parts yellowish with large
brown spots, and towards the belly is golden tinged
with pink, and with large scarlet spots on and below
the lateral line. It has a hard stomach or gizzard, is
partial to a rocky bottom, and may be taken with a
worm or a fly. It attains a weight of 12 lbs. and the
fishermen say that it is a very inferior fish for the
table. Thompson contrasts this form with the Great
Lake Trout, which grows much larger and is
described as a silver-grey with black spots, the
males having a salmon tint below, and the lower
spots enclosed in orange rings; this is taken on
night-lines baited with pollan or perch, and is,
according to Thompson, the common trout of Lough
Neagh.

In a case like this one would like to know whether
there is an incipient species formation due to
physiological isolation; whether 'once a Gillaroo
always a Gillaroo' would hold good, or whether the
larger Gillaroo adopt the habits and assume the
livery of the ferox, thus taking a new lease of life.
Supposing the former to be the case, it would then
be a question whether the two forms keep apart
when breeding, and if so whether the offspring
follow in the footsteps of the parents.

According to Mr. Harvie-Brown the trout of Loch
Mulach Corrie are 'Gillaroo'. Mr. Eagle Clarke tells

me that these are beautiful fish, very deep in the body, and with brilliant red fins.

The feeding in the loch, one may note, is very good, consisting largely of leeches.

The fourth loch surveyed by the B.S. in the Inver basin need not detain us long. Loch Awe is at the head of the River Loanan, at a height of 504 ft. The maximum depth is 8 ft. and the mean is 5 ft. The fishing has been much improved of recent years and it is now a good salmon loch. The principal feeder, the burn flowing from Loch na Gruagaich, enters the loch at its northern end, within 30 yards of the mouth of the River Loanan, which carries the outflow into Loch Assynt.

The next loch that we have to consider is Loch Beannach. From a height of between 230 and 280 ft. it flows into Loch Bad nan Aighean (not sounded) and is 1 ¼ miles long. The maximum depth is 38 ft. and the mean 13 ft. The floor of the loch is very uneven. It falls in four places below the 20-foot level, the deepest part of the loch being in the south western portion, the maximum depth of 38 ft. being observed about 100 ft. to the north of the small island lying off the southern shore.

The two remaining B.S. lochs are not actually in the basin of the River Inver itself, but in that of a small secondary basin which falls into Loch Inver (sea loch) and may be called the Culag basin from the lowest loch in it. Loch Culag is 72 ft. above the sea and is half a mile long, the maximum depth being 9 ft. and the mean 3½ ft.

The last B.S. loch named is called Loch Druim Suardalain and is farther up the little river at a height of 133 ft., opposite Glencanisp Lodge. It is three quarters of a mile in length, and has a maximum depth of 31 ft. and a mean of over 10 ft.

Both lochs are good for salmon and sea trout.

Of other lochs in Assynt, one may name in the first place, on the same Culag River, Loch na Gainimh (550 ft.) between Suilven and Canisp, three quarters of a mile long; and Lochan Fada (700ft.) and a mile long. Also on a tributary of the River Inver, which is more or less parallel to the River Culag, there is the considerable Loch Feith an Leothaid at a height of 860 ft. and about half a mile long; with smaller lochs farther up the burn, e.g. Loch na Faoileige, which once held trout up to 5 lb.

Then, before we come to Inverkirkaig, on another small river falling into the sea west of the River Culag, there is the considerable Loch Bad na Muirichinn, three-quarters of a mile long, and broad, with islands.

Kirkaig

Three miles or so south of Lochinver, the River Kirkaig enters the sea. Three miles upstream, there are falls which salmon cannot ascend, and three-quarters of a mile further up we come to the Fionn Loch, first of a chain, or rather a group, of considerable lochs.

> The great feature of this loch is the existence of alluvial terraces surrounding it. The two lowest are the most extensive, together having an average breadth of 100 yards, their heights about 20 and 30 ft. above the surface of the loch. When the water stood at this level, Loch Fionn must have been connected with Loch Veyatie (the next higher loch), the difference of their levels, as observed by the lake survey, being only about 8 ft. There is another still higher terrace seen to the north. (B.S.).

The head of the Fionn Loch, we may note, lies under the summit of Suilven. Fionn Loch is nearly 2½ miles in length, with a breadth of one-third of a mile. The maximum depth is 90 ft., but the mean depth is only 20½ ft. The height of the loch is 357 ft. above the sea.

The loch contains ordinary trout, char and ferox up to 10 lb. at least.

The next loch, Loch Veyatie, where a 16 lb. ferox
was reported in 1938, is a short way further up the
river, at a height of 366 ft. or, as we have seen,
about 8 ft. above Fionn Loch. Loch Veyatie is four
miles long but, as already indicated, was as much as
7¼ miles at an earlier period. The maximum depth
is 126 ft., the mean depth being 41 ft.

The floor of the loch is uneven, with a few islands
here and there along the shores, and some of the
bays are filled with weeds. The ground around the
loch is low except where Cul Mor rises to a height
of 2700 ft. to the south-west and Suilven (2399 ft.)
to the north-west. In connection with Loch Veyatie,
we have to take into account a complication. A short
way above the foot of the loch, another loch enters it
by a channel which is only 20 ft. in length and in a
flood the two lochs become one. The second loch
has a difficult name - Loch a' Mhadail, which is said
to be pronounced 'Vattle'. It is over half a mile in
length, with a maximum depth of 69 ft. and a mean
depth of 30 ft.:-

> When the water is low the separation between the
> two lochs is complete, the barrier being formed by
> one of the basic dykes so numerous in this part of
> the gneiss: the rock is in places covered by a thin
> layer of sand. The ground around the loch rises
> steeply up to a height of 100 to 200 ft above the
> surface of the water, so that the loch is almost shut
> in, and only towards Loch Veyatie can any
> opening in the wall of rock be seen. (B.S.).

The next loch in the chain is Loch Cam, 409 ft. above the sea and 44 ft. above Loch Veyatie. The length of the loch is 2¾ miles and it has a maximum depth of 122 ft. with a mean depth of nearly 38 ft. The south-eastern portion of the loch is shallow, very few soundings exceeding 20 ft. being recorded. Most of the islands are found in this part of the loch, Eilean na Gartaig being the largest, while Eilean na Gaoithe is remarkable for the long spit of sand and shingle which stretches from its northern point to a distance of nearly 100 yards. The main basin is contained in the north-western portion of the loch where the bottom falls in two places below the 100-foot level. The slope toward the north-west is very steep; in one place a sounding of 90 ft. about 20 ft. from the shore; the cliff above is almost vertical and 50 ft. in height.

We may now pass up to the three lochs that are above Loch Veyatie. Two streams enter that loch at its head. First, there is the main stream called Amhainn Mhor, from the next highest loch, the Cam Loch, which falls by a magnificent cascade into Loch Veyatie. Second, there is a considerable burn, the Amhainn a' Chnocain, that comes from the south, from the Cromalt Hills.

At the head of the Cam Loch we come to a stream which has two tributaries: firstly, the Ledbeg River, from what may be regarded as the fifth Kirkaig loch,

Loch Borralan, at a height of 458 ft., and secondly, the Na Luirgean, from the sixth Kirkaig loch, Loch Urigill, which is a little higher than Borralan, viz. 515 ft. We may begin with the first of these two lochs.

Loch Borralan is over a mile in length and has a maximum depth of 21 ft. and the rather considerable mean depth of 9½ ft. As the B.S. points out 'the abundance of char in the loch is remarkable considering its shallowness. As a matter of fact, it is the shallowest loch in Scotland that is known to have char.'

Mr Tate Regan gives no clue as to the species of char to which he supposes the char of the Kirkaig basin belong, except that, in the number of vertebrae, the Borralan char are to be classed with those of Loch Loyal rather than with those of Scourie and Loch Stack.

Loch Urigill is nearly two miles in length, with a mean breadth of nearly half a mile. It is on the whole very shallow, nearly 99 per cent of the lake floor being covered by less than 25 ft. of water. The deepest part of the lake is near the north-west end, where there is a small central area exceeding 20 ft. in depth, the maximum being 40 ft.

Of the smaller lochs in the basin, much the largest is Loch nan Rac (on the north side of Loch Veyatie), about a mile long; a rather sandy loch with fair trout.

Running into the River Kirkaig below the lower falls is a considerable loch called Loch a' Chapuill, more than a third of a mile long.

One may note that the basin has a comparatively small drainage area in relation to the extent of the six lochs themselves. There are few other lochs in it, the largest being na Tri Lochan which is really a combination of two or three little lochs, but of some extent, nearly a mile long.

Finally, there are many small lochs in the Kirkaig basin, and it is hardly necessary to say that these small lochs should be examined for char as many of them run into char lochs.

Polly

We now come to the basin of the River Polly, a small salmon river which enters the sea three or four miles from the mouth of the River Kirkaig. The primary loch in the basin is called Loch Sionascaig (or Skinaskink as it used to be called). The B.S. informs us that its length is 3⅛ miles, but this does not give us a very clear idea of the loch, which is very irregular and is rather like the shape of a starfish. The area of the loch is considerable for it is the twenty-eighth largest in Scotland. The length of the shoreline of Loch Sionascaig is 17 miles. The height of the loch is 243 ft. above sea level and it is deep (216 ft.).

The lowest two sections of the loch, near its outlet are distinguished as the Polly Lochs. One of them, is called Loch Uidh Tarraigean, clearly with reference to char. The name therefore suggests the existence of char in Loch Sionascaig, and probably implies that it is here that the char of the loch spawn - that is, near the outlet, as we know to be the habit of fish in other lochs, for instance Lochs Killin and Tarff near Inverness.

The B.S. gives two lochs in the basin (besides Loch Sionascaig) which are connected together and eight ft. higher. The smaller one runs into the head of Loch Sionasacaig and is distinguished as Lochan Gainmheich. It is about half a mile across and is 59 ft. deep. Above it, the larger loch, Loch an Doire Dhuibh is a mile long and 120 ft. deep.

The drainage basin of Loch Sionascaig is extremely small and contains very few lochs besides those named. A considerable loch, Loch Doire na h-Airbhe, about a mile long flows into the River Polly below the outlet of Loch Sionascaig, and another loch, Loch a' Ghille, of some length, enters the main loch on its north side. These lochs alone need to be mentioned, except perhaps four very small lochs on the south slope of Cul Mor which are on a burn running into Lochan Gainmheich, two of them at a height of 1000 ft.

Garvie

Loch Garvie, which gives its name to this basin, is a very small loch about quarter of a mile long at the outlet of the basin into an inlet of Enard Bay. Only the three main lochs have been measured by the B.S. These form a connected series

The uppermost loch, Loch Lurgain, flows through a small fourth loch, Loch Bada na h-Achlaise (or the Green Loch), thence, almost directly connected, to Loch Bad a' Ghaill by the Abhainn Osgaig into Loch Osgaig, thence through Loch Garvie into the sea. The second in the chain, Loch Bada na h-Achlaise, was not sounded. Loch Osgaig, whose maximum depth is 153 ft. and mean depth 47 ft., is at a height of 57 ft. and has a length of two miles. Loch Bad a' Ghaill, which is taken as on a level with the Green Loch, is also two miles in length, having a maximum depth of 180 ft. and a mean of 62 ft. Its height is 173 ft. - 100 ft. above Loch Osgaig. Loch Lurgain, the largest loch in the chain, is four miles in length, has a maximum depth of 156 ft. and a mean depth of 61 ft. Its height is also 173 ft. above sea level.

VII

BROOM AND THE GRUINARDS

Broom

On the next section of the west coast - from the basins of the Garvie to the Gruinards - we draw a blank as far as char are concerned. Five basins can be distinguished, those of the Rivers Achnahaird, Kanaird, Ullapool, Broom and Dundonnell.

The basin entering Achnahaird Bay has several lochs, two of them quite large, Loch Vatachan, running into Loch Raa - but not surveyed by the B.S.

The River Kanaird flows into the sea-loch, Loch Kanaird. None of its, mostly small, lochs were surveyed by the B.S.

The basin of the Ullapool River presents nothing of interest. Loch Achall contains salmon and sea trout and is moderately deep (70 ft.). There are a few small lochs in the district, but no record of char in them, nor in Loch Achall itself.

The basin of the River Broom is equally without interest for our purpose. It has two main tributaries - the Abhainn Cuileig from Loch a' Bhraoin (Broom) and the Abhainn Droma from Loch Droma, near the watershed between the Broom basin and that of the Blackwater (which joins the River Conan near Contin). Loch Droma is shallow, but Loch a' Bhraoin is fairly deep - deep enough for char - and it is fairly high (over 800 ft.). There is, however, no record of char.

The Dundonnell River is a small salmon and sea trout river - it need not detain us.

Gruinards

At the River Gruinard, we strike char again. Only one loch is given in the B.S., viz. Loch na Shellag.

The River Gruinard is a good salmon river which rises in Loch an Nid at a height of over 800 ft., presumably in the mediaeval forest of the Nest. After a further course of nearly six miles, it enters Loch na Sealga (Shellag, the loch of the huntings). Five or six miles below the river falls into Gruinard Bay (= Grunn-fiord, the narrow fiord).

The main body of Loch na Sealga forms a simple basin, the deeper water occupying the wide south-eastern portion, while the water shoals gradually as the loch narrows in outline towards the north-west. The lower very narrow portion of the loch is less than 50 ft. in depth, except for a small subsidiary basin situated about a quarter of a mile from the outflow in which depths of 52 to 56 ft. were recorded. The slope of the bottom is thus in striking contrast at the two ends of the loch, for while a depth of 100 ft. may be found about 250 ft. off the south-eastern end, where the Abhainn Srath na Sealga flows in, one must proceed more than a mile from the opposite end, where the Gruinard River flows out, before encountering depths exceeding 100 ft. The 100-foot basin is nearly 2¾ miles in length and the main 200-foot basin is over three quarters of a mile in length.

An inspection of the map shows that the deeper part of the loch is distinctly flat-bottomed in character.

The temperature observations taken on August 13 and 14, 1902, are interesting:

Depth	°F.
Surface	55.0°
30 ft.	53.0°
50 ft.	53.0°
75 ft.	51.4°
100 ft.	48.5°
180 ft.	47.9°

The head of Loch na Sealga lies between An Teallach (3483 ft.) which is well known to rock climbers, and Beinn Dearg (2974 ft.). At a height of 278 ft. the loch is 3¾ miles long, and trends in a south-east and north-west direction. The maximum

depth is 217 ft. and the mean depth 103½ ft. The
loch is very good for fishing and contains salmon,
sea trout, ferox and char in abundance. The river
running out of the loch is known as the Great
Gruinard or simply as the Gruinard, by contrast with
the Little Gruinard, which also enters Gruinard Bay.

The loch is deep and is well known to contain char.
They are said to spawn in the Abhainn Gleann na
Muice, a burn which enters the main river near
Larachantivore, fully half a mile above the loch, and
are usually about nine or ten inches in length. They
are greenish-purple on the back, with pale reddish
spots, and reddish on the belly, and take freely.

The basin holds several smaller lochs which may
have char. The largest, Loch an Eich Dhuibh, about
a mile in length and 737 ft. above the sea, flows into
Loch Mor Bad an Ducharaich, the outflow from
which runs into the River Gruinard a short distance
below the foot of Loch na Sealga. On another burn
which joins the main river on the same bank further
down is Loch Gaineamhaich, nearly three quarters
of a mile long, and less than 700 ft. above the sea.

Near the mouth of the River Gruinard, on the same
bank, a tributary enters from Lochan na Cairill,
which is nearly three-quarters of a mile long, but not
high (under 400 ft.). On the left bank of the river,
nearly opposite the highest burn on the right bank, a
burn comes in from three lochs, of which the

longest, Loch Ghiubhsachain, is three-quarters of a mile in length. One of two other lochs which run into it is nearly 1750 ft. above the sea.

Finally - and most likely for char - there are three small lochs, very high up (about 2000 ft.) on the burn which is chiefly used for spawning by the char of Loch na Sealga. This burn enters the main river, as we have seen, on the left bank, above Loch na Sealga.

We may now pass the small Inverianvie River, which enters Gruinard Bay between the two Gruinards and has two considerable lochs on it. Loch Toll a' Mhadaidh, at its source, is 1345 ft. high, and three quarters of a mile long. Loch a' Mhadaidh Mor, nearly three quarters of a mile long, is two miles from the foot of the river, above a fall, about 600 ft. above the sea, with two small lochs running into it.

Next, we come to the Little Gruinard River, in whose basin is included the Fionn Loch and three other lochs given in the B.S.

The Little Gruinard River rises in a well known loch, the Fionn Loch, which is 559 ft. above the sea and has a greater area than Loch Assynt. The Fionn Loch (without the Dubh Loch) is 5¾ miles long, the maximum breadth being 1½ miles. The maximum depth of 144 ft. was observed in two places, one near the south-eastern end, the other in the central part opposite the entrance of the stream bearing the outflow from Lochan Beannach. The mean depth is 57¾ ft.

The bottom of the loch is most irregular and the contour lines most sinuous in character; the north-western end is filled with boulders, which often rise out of comparatively deep water in an astonishing manner. The main 50-foot basin is nearly four miles in length, approaching quite close to the south-eastern end, and extending between the islands called Eilean Fraoch and Eilean an Eich Bhain. A second 50-foot basin runs in a north and south direction along the centre of the large arm thrown out in a northerly direction towards the foot of the loch, extending to the west of Eilean an Eich Bhain, and is nearly two miles in length. A third small 50-foot basin occupies the extreme north-western end of the loch.

> The head of the loch is practically continuous with the Dubh Loch which is separated only by an artificial causeway built on a sandbank. When the water is high the causeway is flooded, though under ordinary circumstances the difference of level is about a foot. The matter was the subject of litigation in 1877, the Lord Ordinary deciding that the lochs were one, but the House of Lords reversed this decision.

Salmon enter the Fionn Loch, but not in great
numbers as there is a fall on the Little Gruinard
River, which runs out of the loch, and they are
hardly ever caught. There are many trout, but
apparently no char. The loch has a great reputation
for ferox, but the facts seem hardly to justify it.
Osgood H. Mackenzie, whose memory goes back to
the time when ospreys built on Loch Maree, gives,
in his *A Hundred Years in the Highlands*, some very
interesting records of fishing in the Fionn Loch.

The earliest record is of Sir Alexander Gordon
Cumming who fished the loch in March and April of
1851. His best day was seven fish - 14½, 12½, 12¼,
12, 10, 6¾, 6½ lbs. - but there is no mention of his
getting any single fish larger than the 14½ lb.
specimen. Mackenzie writes:

> My uncle happened to come across the late Sir
> Alexander Gordon Cumming of Altyre, who was
> then a very keen young sportsman, and on hearing
> of the reported size of the trout, Sir Alexander
> determined to try the loch himself. Of all unlikely
> times of the year for trout fishing, he chose the
> middle of March, when no one but himself would
> have hopes of catching anything, but in spite of the
> odds against him, he caught plenty of fish, some of
> which were real giants.
>
> The old people declared there were three different
> species (or at least varieties) of these big trout and
> gave them three different Gaelic names, viz.

Claigionnaich (meaning skully, big-headed), Carraigeanaich (stumpy, short and thick) and Cnaimhaich (bony, big-boned).

The distinction between the first two seems simply that between a ferox out of condition and one in good condition, while the meaning of the third type is not clear.

Next, the experiment was tried of setting a lythe net in the loch:-

Just once (perhaps about the year 1863) I set a net in the Fionn Loch, and got such a haul of fish that the two men who went to lift it could hardly carry them home across the moor. The biggest of the lot scaled eighteen pounds, and I have heard of one other having been caught of similar weight.

A detailed record of baskets taken by Mr F.C. Grady from May 28 to July 26 in 1912 is given. The total of these baskets was 3625 trout weighing 1410 lbs. and this included 9 ferox, the largest of which was 8¼ lb. The fly trout seem to have weighed more than three to the pound.

Lastly, of a fish which O.H. Mackenzie lost 'a few years ago' (his book was written in 1921), both he and his stalker were sure that this was quite 'double the size of any that we had ever seen before. It jumped three times clean out of the water close to the boat and we saw it as well as if we had handled it. Without in the least wishing to exaggerate, I honestly declare that fish to have been a twenty five pounder.' Clearly we may assent to the estimate.

The Dubh Loch is situated in a corrie forming the natural head of the Fionn Loch, surrounded by high and steep hills. The loch is about 1¼ miles in length. The maximum depth of the loch is 88 ft. and the mean depth is 42 ft.

The conformation is quite a contrast to that of the adjoining Fionn Loch. Except for one or two irregularities of the lake floor, Dubh Loch forms a simple basin, all the contour lines, though showing occasional sinuosities enclose continuous areas.

We may now cut across the Rudha Mor between the Little Gruinard River and Aultbea, passing five small river basins which contain among them three lochs given by the B.S. (two classed under the Gruinard basin and one in the Ewe basin).

The Allt Bad an Luig is a small stream which enters Gruinard Bay fully a mile west of the Little Gruinard River. Only one loch in it has been measured by the B.S., but three others should be named as they are of considerable size and may be of more importance for fishing.

Loch Fada, four miles up the stream and at a height of 498 ft. is 1½ miles long, and has a maximum depth of 56 ft. and a mean depth of over 17 ft. Of the entire lake floor 45 per cent is covered by less than 25 ft. of water.

Loch Moine Sheilg, which is half a mile long, is connected with the main stream by a small burn which joins it near its foot.

On the main stream, about two miles up it, and above Loch Fada, is the considerable Loch na h-Uidh. Into this loch, there enters a burn from a small loch, which is, in turn, fed by small lochs higher up.

Finally, at the head of Loch Fada, a small burn runs in from Loch an Iasgair (an auspicious name), which is nearly half a mile long.

The lochs in the Rubha Mor, the point that projects between Gruinard Bay and Loch Ewe, were not completely stocked till 1907. Nevertheless two, which presumably had native trout, were surveyed in 1902.

Loch na Beiste ('of the monster') – to reveal which, in 1870, the loch was once drained – is at a height of 168 ft. on a small burn which is a tributary of the stream that enters Gruinard Bay at Udrigle. It is over one third of a mile in length, has a maximum depth of 35 ft. with a mean depth of 10½ ft. It has good trout, between a half and three quarters of a pound.

Loch an t-Slagain is on a small stream running into
Slaggan Bay at the entrance of Loch Ewe. It is two
thirds of a mile in length and is at an elevation of
103 ft. The maximum depth is 55 ft. (very near the
south-eastern shore) and the mean depth is 16½ ft.

We may obviously pass Loch nan Dailthean, which
outflows to the sea Loch Thurnaig (an inlet of Loch
Ewe) which is 'so shallow that cows may walk over
the whole of it.' Similarly, we come to several small
basins, all with little known lochs - Laide,
Achagarve and Opinan basins - before Greenstone
Point and then the Slaggan basin.

We come then to the basin of the Allt Bheithe at
Aultbea, which has two lochs in the B.S. The chief
loch is Loch a' Bhaid-luachraich, which is nearly
two miles up the burn, is about a mile and a half
long, at a height of a little over 300 ft. It is
surrounded by low rounded hills, steep only towards
the south-east, covered with peat or morainic
material. It is extremely irregular in outline and in
fact may almost be looked upon as two lochs with a
connecting arm. The south-western part is shallow,
the maximum depth observed in it being 143 ft., the
mean depth is 34 ft. The maximum depth in the
connecting arm is 15 ft. The lower part, where the
stream leaves the loch, is thickly overgrown with
reeds and rushes. The loch drains an area of 3¼
square miles and contains ferox and char, which are
said to spawn near the outlet of the loch.

Temperatures observed on June 29, 1902, were:

Depth	°F.
Surface	55.4°
30 ft.	55.4°
50 ft.	53.6°
75 ft.	49.8°
85 ft.	48.4°
120 ft.	47.2°

The other loch given in the B.S., Loch Mhic Ille Riabhaich, is nearly 300 ft. higher and runs into the main loch. Presumably therefore it may hold char, although it is not deep.

Two smaller lochs run into the lower loch, and three very small lochs into the upper one.

VIII

EWE, GAIRLOCH, TORRIDON, CARRON AND ALSH

The interesting and important basins of Ewe and Gairloch have already been dealt with in Part I of *Ferox and Char*. Three of the lochs there are known to contain char - Lochs Maree, Clair and Coulin.

Torridon

Resuming our survey of river basins on the west side of Gairloch, we proceed around the Red Point, and passing a few miles of coast which present nothing of interest, we reach the small Craig River, on which there are four or five lochs to the west of Ben Aligin. The first we come to is Loch Sgeireach at a height of 821 ft. and nearly half a mile long. Then two small lochs at nearly 800 ft. and after that a considerable loch, Loch na h-Uamhaig, half a mile long and at a height of nearly 1000 ft. Last on this river, a smallish loch over 1000 ft. high.

Returning to the shore, near the head of Loch Torridon, we are at the sea loch, Loch Diabaig, a natural harbour north of Loch Shieldaig on the south side. High up, 443 ft. above this natural harbour, is a double loch more than a mile in length. It was stocked some years ago (perhaps twenty) and has some fair trout, the upper part, Loch Diabaigs'

Airde, being the better. Five or more very small lochs drain into it.

Four basins drain into Upper Loch Torridon. Three of them are small but are perhaps worth noting. First, on the north shore, near Torridon House, there is a tributary of the Amhainn Coire Mhic Nobuil, from one of the many 'lochs of the monsters', with falls on the way. It is nearly half a mile long, and very high - just north of the Sgurr Mhor of Ben Aligin. The little 'river' itself rises in two or three small lochs at over 1000 ft, close to the summit of Liathach, and little more than a mile from the loch in the famous corry, Coire Mhic Fhearchair.

The second river basin is that of what is called the River Torridon which falls into the head of Upper Loch Torridon, along with the third river which is called the Abhainn Thrail. The former rises in a little loch, Lochan an Iasgaich, which has salmon and sea trout, while the latter has its sources in the Ben Damph Forest, chiefly in four lochs in sequence at heights of over 1000 ft. above a waterfall. The second last of the series, Loch an Eoin, is half a mile long and fairly broad.

This brings us to the fourth river basin that falls into Upper Loch Torridon, that of the River Balgy, and we again come in touch with the B.S. One loch has been measured in the Balgy basin, namely Loch Damh, which is the source of the river. The loch is only 129 ft. above the sea, so that the river, with a fall on it, is short - only about a mile long. The length of the loch is about four miles, and going up it from the foot we find that it first trends towards the north-west and then towards the south-west. The maximum depth of 206 ft. was observed at the widest part of the loch, at the junction of the two limbs of the V, and the mean depth is 59 ft.

Loch Damh is complex in conformation, there being three basins separated from each other by shallower water. The northern basin is unimportant, with a maximum depth of 34 ft. The southern basin is of simple form, over three-quarters of a mile in length and encloses a 100-foot basin over half a mile in length, having a maximum depth of 135 ft. The central basin is the largest and deepest, being two and a half miles in length, and enclosing towards the southern end a small 200-foot basin elliptical in form and nearly a quarter of a mile in length. At the northern end of this central basin, the 50-foot contour is irregular, shallow water extending towards the middle of the loch, and approaching very close to a small area slightly exceeding 100 ft. in depth. Here, a sounding of 42 ft. was recorded less than 30 yards from one of 115 ft., giving a gradient of nearly one in one. The shore slopes to the east of the 200-foot basin are fairly steep.

Temperatures taken in the deepest part of the loch on August 21, 1902, were:

Depth	°F.
Surface	56.5°
50 ft.	56.0°
100 ft.	48.5°
150 ft.	43.1°

The sprungschicht [thermocline] is between 50 and 100 ft.

Temperatures in the southern basin on August 22, 1902, were as follows:

Depth	10.45 am °F.	5.30 pm
Surface	56.8°	56.6°
30 ft.	56.2°	-
50 ft.	55.1°	56.0°
100 ft.	48.1°	48.5°
120 ft.	48.0°	-
150 ft.	-	43.1°

There are several subsidiary lochs.

A few feet above Loch Damh, after half a mile of river, there is a double loch, or rather perhaps two lochs, Loch Coultrie and Loch an Loin, on the same level. They were not surveyed and were seen to be largely filled with weeds, and are apparently shallow.

Two small streams which join to enter Loch Damh near its head come from hill lochs, one from Loch Coire an Ruadh-staic, which lies between the conspicuous hills, Maol Chean-dearg (3060 ft.) and An Ruadh-stac (2919 ft.), which is nearly a third of a mile long and almost 1500 ft. high; and the other from two smaller lochs, the higher of which is over 1250 ft. high.

As to fishing on the River Balgy and the three lochs (Loch Damh, Loch Coultrie and Loch an Loin), there are four beats on the Balgy, but neither salmon nor sea trout are got above the falls pool or in Loch Damh. But they are got in the two upper lochs (Loch Coultrie and Loch an Loin). Ferox up to 9 lb. have been got in Loch Damh and ordinary trout are taken with fly up to the usual 14 ounces.

A char was caught in Loch Damh many years ago (1927), thought to have been put in by the Duke of Leeds.

The small Loch Dhughaill has also been measured. It lies in a small separate basin with Ben Shieldaig between it and Loch Damh, and near the village of Shieldaig. It is about half a mile up a stream, the Abhainn nan Lub, that runs into the head of Loch Shieldaig, a branch of Loch Torridon and 84 ft.

above the sea. Its length is more than half a mile, and the maximum depth is 108 ft., with a mean depth of 38¼ ft.

The loch holds sea trout and a few salmon.

Having accounted for the small valley of Glen Shieldaig, we are now approaching the district of Applecross which is a point on the coast between Loch Torridon (the lower loch) and Loch Carron, or, more exactly, Loch Kishorn. As a point, it presents the usual perplexity of a multiplicity of small rivers and small basins. In the case of Applecross this difficulty is aggravated, for it is a peninsula rather than a simple point.

One, however, of these many rivers is conspicuously indicated by the name itself, for Applecross is a corruption of Abercrossan, which in Pictish means the estuary or mouth of the River Crossan. This, as it happens, is not merely one of the rivers in the peninsula, but is the longest.

Here, at Applecross, Maelrupha, the associate of St. Columba, landed and built a church which he made the centre of a sanctuary. Any fugitive who should win to within nine miles of the church, that is, to the peninsula of Applecross, was safe. It will perhaps be remembered how R.L. Stevenson makes effective use of the word to suggest the wildest and most remote region on the mainland of Scotland.

Starting then from Glen Shieldaig, we come first to
the 'white estuary' (Inverbain), so called, no doubt
from the waterfalls on the river. Three miles or so
up the Amhainn Dubh, at a height of nearly 800 ft.,
is Loch Lundie, which is nearly two miles long and
much the largest loch on the peninsula (not
surveyed).

Loch Lundie, as well as two small lochs half a mile
down the glen had no trout till they were stocked by
Donald Matheson, thirty years ago (or more). All
three now have good trout.

At Inverbain, a path, six miles in length and about
1200 ft. high, crosses the peninsula to Applecross.
From Inverbain, the coast begins to trend towards
the north-west until it rounds the extreme north
point of Applecross, when it begins to turn
southwards for about two miles to a place called
Cuaig. In the first stretch of coast after passing
Inverbain, where it runs north-westwards and in the
further two miles where it runs south to Cuaig, there
are some six or seven small river basins which
collectively yield about fifteen small or smallish
lochs, up to a third of a mile or so in length.

At Cuaig there enters the sea a longer stream which
comes from the considerable Loch Gaineamhach,
which is rather irregular in form, and may be
described as nearly three-quarters of a mile in

length; near it, is the small Loch na h-Eargaich, which is one of the lochs called a 'night loch' in the Highlands (see page 211 in Part I of *Ferox and Char*).

Between Cuaig and Applecross, where the coast runs almost due south, there are four river basins, but the lochs in them are very small except two. An Dubh-loch (1229 ft.) is half a mile long, and Loch nan Eun (over 1000 ft.) is three-quarters of a mile in length. Another loch runs into it from a height of nearly 1750 ft.

The River Applecross rises in a loch in the Coire Attadale, one of the fine corries in the deer forest of Applecross. The loch, Loch Coire Attadale, is nearly half a mile long, and over 1000 ft. high. Another small loch runs into it. A tributary to the River Applecross from the north has two smallish lochs on it. Another tributary, this time from the south, has a number of very small lochs on or near it.

After passing Applecross, we are beginning to round the south point of the peninsula and in the end we are going eastwards and even north-east. Here there are six small basins with lochs in them, but only two lochs need be mentioned, Loch Braigh an Achaidh, nearly three-quarters of a mile at a height of almost 1000 ft., with small lochs running into it; and Loch Coire nan Arr, half a mile long and exiting via the Russel Burn into Loch Kishorn.

Finally (in Applecross, at the limit of the peninsula and of the old sanctuary) we have the River Kishorn which rises in another Loch Gaineamhach, fully half a mile long, at a height of over 1000 ft. and with two small lochs running into it. It is said to hold fine trout, up to four or five pounds, but there is no mention of char. It lies just to the west of one of the great Applecross corries.

Carron

We have next to describe the lochs in the basin of the western River Carron. There are not many in it, and only two of them have been surveyed.

The more important is a second Loch Dhugaill, larger than the loch in Glen Shieldaig. It is more than four miles up the river, at a height of 93 ft. The loch is two miles in length, but it tapers off towards the west, the lower end, for half a mile being merely a series of small expansions of the river. It is surrounded by lofty mountains, Fuar Tholl (2975 ft.) rising to the north-west and Creag a' Chaoruinn Eagan (2260 ft.) to the south. On the south side also, the land rises steeply to the ridge of Creag an Eilein, the highest part of which is about a quarter of a mile distant. The maximum depth of 179 ft. was observed here less than half a mile from the head of the loch.

The main body of the loch is simple in
conformation, the contours following approximately
the shore lines, and the 100-foot basin is three-
quarters of a mile in length. There are two small
basins over 25 ft. in depth in the river expansions at
the south-western end of the loch.

Temperatures were measured at 4.30 pm on August
7, 1902:

Depth	°F.
Surface	54.0°
60 ft.	53.5°
70 ft.	50.0°

A fall of 3.5° between 60 and 70 ft. indicates the
sprungschicht, the decrease both above and below
being gradual.

The other loch that has been measured is Loch
Sgamhain. It lies about six miles farther up the
River Carron, near the head of the glen, with
Moruisg (3026 ft.) to the south, and at a height of
491 ft. The distance between it and Loch Gowan, at
the head of Strathbran (in the basin of the River
Conan) is only about 2½ miles and Loch Sgamhain
is only 23 ft. lower.

Loch Sgamhain is more than a mile long and has a
maximum depth of 72 ft. and a mean depth of nearly

27 ft. The 25-foot basin extends nearly from end to end of the loch, but is very narrow in the western portion. The wide eastern basin includes the 50-foot basin which occupies a central position and is about one-third of a mile in length.

Both lochs have salmon, sea trout, ferox, ordinary trout and char. The S.G. says of Loch Dughaill that ferox weigh 'up to 8 lb. and char up to 2½ lb.'

There are several subsidiary lochs. First, going up the right bank from the foot, we come to the Amhainn Bruachaig, a burn which comes down from the high Loch a' Mhuilinn (over 1500 ft.).

Second, the Fionn-amhainn from three small lochs, two of them at about 1500 ft. on a tributary from the An Ruadh-stac, and a third from a larger Loch Coire Fionnaraich (one third of a mile in length).

Third, and most interesting, Loch Coire Lair, nearly half a mile long and high (1259 ft.) on the River Lair. The S.G. says that this loch 'is about 1¾ miles round and contains char in great abundance. They run small, take fly freely, and as many as ten dozen can be got in a day.' The statement rather implies that there are only char in the loch. If so, the loch probably had originally no fish, and was then stocked with char only. There is a small loch called Loch a' Bhealaich Mhor on the pass between Sgorr

Ruadh (3142 ft.) and Fuar Tholl (2975 ft.) running
into it from a greater height (2000 ft.).

Farther up the right bank, only one small loch,
Lochan Meallan Mhic Iamhair, runs into the River
Carron just below where it leaves Loch Sgamhain.
On the left bank there is only one considerable loch
near the foot of the river at Strathcarron, the double
Loch nan Creadha, with three smaller lochs running
into it, in sequence.

Alsh

We next go round the point between Loch Carron,
Loch Alsh and Loch Long.

First, on a tributary of the small River Attadale,
which enters Loch Carron a mile below the head, is
a loch half a mile in length with the promising name
of Loch Iasaich, at a height of 1000 ft.

In another small basin, two miles further down there
is Loch na Sroine (over 1000 ft.) with a still smaller
loch entering it.

Then we come to Fernaig with the Allt Cadh an Eas
in Strath Ascaig, whose basin has two considerable
lochs in it: first, Loch nan Gillean, more than half a
mile long, and second, running into it, Loch na
Leitire.

Next, at Duirinish, there is another small basin, the Allt Duirinish with Loch Lundie on a tributary and Loch Achad na h-Inich at the source. The latter is half a mile long, and as broad.

We have now passed the Kyle of Lochalsh and our next basin is that of the Balmacara Burn, which, however, contains only small lochs with small trout.

At last we come to a basin, that of the Allt Gleann Udalain, flowing into Loch Alsh east of Balmacara, which contains a measured loch. There are two other considerable lochs on the west side of the basin, Loch na Smeoraich (902 ft.) and, running into it Loch a' Ghlinne Duirch. Slightly further east in the basin is Loch Anna (measured) and Loch na Gaibhre. The former is at a height of 1040 ft., and a third of a mile in length, and runs into the latter whose maximum depth is only 27 ft., and whose mean depth is 13 ft. It has a drainage area of only half a square mile.

This takes us to the beginning of Loch Long, a branch northwards of Loch Alsh, and Loch Duich, which runs towards the south-east. Going up the west side of Loch Long, and beyond the mouth of the River Ling, we come first to the small basin of the Allt Dearg, a tributary of the River Ling. Of the small loch on this tributary, Loch na h-Onaich, about a third of a mile in length, Professor W.J.

Watson has the following note: 'Gaelic, indicating Place of Foam'.

On another tributary further up the main stream is a loch of similar size, Loch Innis nan Seangan. A tributary on the other side entering below the Allt Dearg comes from Loch Beinn a' Mheadhoin.

There is nothing of interest in the upper valley of the River Ling until we reach its source in Loch an Laoigh, which is very nearly a mile long and at the considerable height of 877 ft. This has brought us within two miles of Loch Calavie in the Beauly system to the east.

Coming down the east side of the River Ling we pass the small Lochan Annie which has yielded trout up to 6 lb., and Loch Cruoshie, half a mile long. We now return to the head of Loch Long and enter the basin of the River Elchaig, which enters the head of Loch Long from the east at Killilan.

Two lochs in the basin of the River Elchaig are described by the B.S. First, five or six miles up the river, there is Loch na Leitreach, more than a mile long, but very narrow towards the foot. It has salmon and sea trout. The height of the loch is 275 ft. The formation of the main basin is simple, the deeper water occupying the wide upper portion of the loch, a depth of 65 ft. having been recorded close to the upper end, and the maximum depth of

88 ft. about half a mile down the loch. The mean depth is over 40 ft.

The other measured loch is a kind of double loch, consisting of Loch a' Bhealaich and Loch Gaorsaic; the former being described by the B.S. It is two-thirds of a mile long and situated at a height of 1243 ft. on a tributary, the Allt an Ghlomaich, three miles or so above the falls, and perhaps four miles above where it joins the River Elchaig. The maximum depth of 44 ft. was found towards the lower (northern) end of the loch, in the vicinity of the two islands. The mean depth is about 16½ ft., and the drainage area is nearly two square miles.

Neither of the last two lochs is likely to have char and there is no record of the fish there.

There are two other subsidiary lochs that should be mentioned

On a tributary that enters the River Elchaig a mile and a half below the Glomach, is Loch nan Eun, a very irregularly shaped loch nearly three-quarters of a mile long, with one or two small lochs running into it. It is over 1250 ft. above sea level.

The source of the River Elchaig (near Loch Lungard in the Beauly basin) is Loch Mhoicean. This loch is nearly quarter of a mile long and very high, almost 1250 ft.

Of the coast between Loch Alsh and Loch Nevis
(opposite Skye) the B.S. remarks that 'the large area
draining into Loch Duich is almost entirely devoid
of lakes' and the same is true of the region that we
have now to consider.

The first basin, that of the Glenmore River, which
falls into Glenelg Bay, has only a small Loch a'
Mhuilinn with a larger Loch Beinn a' Cha-oinich
running into it. On the other hand, the Glenelg
River, just south of Glenelg, has on one of its
tributaries, one of the mysterious 'night lochs' - Loch
Bealach na h-Oidhche - at a considerable height
(nearly 2000 ft.) - see page 211 in Part I of *Ferox
and Char*. It is less than a mile from Beinn
Sgritheall, which impends over Loch Hourn.

Before we reach Loch Hourn, the Allt Mor Shantaig
comes out of a considerable loch of very irregular
shape - Loch na Lochain, at a height of almost 1100
ft.

Entering Loch Hourn, we come to the River
Arnisdale, on which, in Gleann Dubh Lochain, we
find a pair of lochs - the Dubh Lochain.

At the head of Loch Hourn, we find the sea loch,
Loch Beag; the burn coming in here, at Kinloch
Hourn, has two branches, that from the south drains
from Loch Coire Shubh with, above it, another

small loch which is less than a mile from Loch
Coire nan Cramh, which drains east to Loch Quoich.

Turning west again, along the Knoidart peninsula,
there is little of special interest until after we have
rounded the point opposite Skye and come back east
again into Loch Nevis. Here, on the Inverie River is
Loch an Dubh-Lochain, which contains char as well
as trout. It is over a mile long, but has not been
surveyed.

There is little else of interest in Loch Nevis and so
we come back west again and out into the Sound of
Sleet, past Mallaig.

IX

MORAR, NAN UAMH, AILORT AND SHIEL

Morar

It seems certain that at one time the outlet of Loch Morar was to the south-west, because the col there does not rise more than 100 ft. above the sea, and there is a narrow belt of comparatively flat ground running southwards towards the source of the burn called Allt Cam Carach. It will be observed by an examination of the B.S. depth map, that the deep water at the west end of the loch runs in the direction of this flat ground.

In other words, according to the B.S., at some previous time, when the loch stood 100 ft. or so above its present level, it was 'tapped' by the River Morar and lowered by 100 ft., and the outflow diverted so that it ran at a height of 30½ ft. by its present channel down the River Morar:-

> Loch Morar is a glen lake which lies in a transverse valley – that it to say, in a valley the direction of which is independent of the geological structure of the region and crosses irregularly the strike of the rocks. This fact probably accounts for the steep sides and the great depth to which the valley has been scooped out. By some observers it is held that the great depth of Loch Morar

precludes the idea that it was scooped out by river
action or by ice. (B.S.).

Loch Morar is 11½ miles long and, as we have seen,
at a height of 30½ ft. above the sea. It is the fifth
largest loch in Scotland, but in respect of depth it is
easily first, its maximum depth being 1017 ft.,
though the mean depth is only 284 ft. (compared
with 343 ft. in Loch Ness). 2784 acres of the lake
floor (or 42.2 per cent) is covered by less than 100
ft. of water. The volume of the loch is only 81,000
cubic ft. as compared with 260,000 cubic ft. for
Loch Ness.

There are several islands, more or less covered with
vegetation, at the west end of the loch, and parts of
the surrounding land are fairly well wooded, but as
one proceeds eastwards towards the head of the
loch, the scenery becomes wilder, the vegetation
more scanty, and the mountains on both sides of the
loch rise higher and more steeply. At many places
on the north shore they rise precipitously from the
water's edge, and around the head of the loch they
reach a height of fully 3000 ft.

In the sea to the west of Morar, there is no depth
approaching 1000 ft., and to get a depth of 1000 ft.
one has to go west of St Kilda and Ireland. There are
no depths comparable to this in the North Sea, but
the submarine valley, known as the 'Norwegian
Gut', which runs round the west and south coast of

Norway, is remarkably deep, the depth of 1794 ft. having been found in that part called 'the sleeve'.

The four deepest lakes in Europe are in Norway and after them come:

Lake Como	1341 ft.
Lake Maggiore	1220 ft.
Lake Guarda	1135 ft.
Loch Morar	1017 ft.

The Lake of Geneva has a maximum depth of 1014 ft., and its height above the sea is 1220 ft. None of the four deep Norwegian lakes is more than 607 ft. above the sea.

Loch Morar is of simple formation, the bottom falling on all sides to the deepest part, but with here and there a few minor undulations of the lake-floor, especially in the wider western half of the loch, where the contour-lines of depth are much more sinuous than in the narrower eastern half. In many places, the contour lines of depth approach each other very closely, indicating that the slopes are very steep in these places. As to the configuration of the bottom of the loch, it is enough for our special purpose to decide only the three uppermost layers of water in the loch.

The area over 300 ft. in depth is 9½ miles in length, extending from 1½ miles from the west end to a

little over half a mile from the east end of the loch.
The area enclosed by the 200-foot contour is nearly
10½ miles in length, extending from about a mile
from the west end to about one-sixth of a mile from
the east end of the loch. The area enclosed by the
100-foot contour extends to over 11 miles in length,
extending from one-fifth of a mile from the west end
of the loch to a short distance from the east end.

The temperatures off Wester Swordland and Meoble
on July 2, 1902, were:

Depth	°F.
Surface	55.2°
10 ft.	54.5°
50 ft.	51.6°
60 ft.	49.1°
75 ft.	48.0°
78 ft.	47.0°
90 ft.	-
100 ft.	-
150 ft.	45.0°
250 ft.	43.3°

In comparison, those taken east of the islands on
July 3, 1902, were:

Depth	°F.
Surface	54.6°
30 ft.	53.0°
50 ft.	50.5°
100 ft.	47.3°
200 ft.	44.2°
600 ft.	42.3°

In *Notes on the biology of Loch Morar* by James Murray, the B.S. has:

> Salmon, sea-trout and loch trout abound in Loch Morar and the sport is frequently good, but the salmon as a rule are 'dour' to rise. Charr and powan or freshwater herring (*Coregonus*) are said to inhabit the loch.

The last sentence is vague and seems unsupported by evidence. [Editors note: It is now known that charr have been recorded from Loch Morar but it is highly unlikely that *Coregonus* occur there.]

There are several subsidiary lochs in the Morar basin

We will begin with the two lochs which are described in the B.S., both of which run into Loch Morar itself. Loch Beoraid, the larger and more important, is a long narrow loch, lying amidst wild and rocky scenery, about three miles to the south of

Loch Morar, at a height of 168 ft. above the sea. The River Meoble, which drains the loch, issues at the west end, and, after a course of three miles, falls into Loch Morar, forming a waterfall over rocks a short distance from its exit. The length of Loch Beoraid is 3½ miles and it has a maximum depth of over 72 ft.

There are two basins over 100 ft. in depth; one at the west end of the loch, three quarters of a mile in length. The maximum depth obtained in it was 147 ft., very near the outflow. The eastern basin is nearly a mile and a quarter in length, with a maximum depth of 159 ft., the area over 150 ft. in depth being almost half a mile in length. Loch Beoraid is a rock basin, divided into two separate basins by a rocky ridge which crosses the loch at the large island. The 50-foot area is continuous from end to end of the loch, passing to the south of the large island, the depth in the channel being 53 ft.

There are two small lochs running into Loch Beoraid: Lochan a' Ghobhain, near the foot and a (larger) Lochan Tain Mhic Dhughaill, a third of a mile long, at a height of over 1000 ft., nearer the head.

The other measured loch in the Morar basin is Loch an Nostarie, which lies to the north of the west end of Loch Morar and drains into it through two smaller lochs (Loch a' Mheadhoin and Loch a' Bhada Dharaich), and the Allt an Loin. It is at a height of 89 ft. and has a length of a little over half a mile, the maximum depth being 35 ft. and the mean depth very nearly 11 ft.

One or two smaller lochs in the Morar basin seem worthy of mention. First, half a mile south of Loch an Nostarie, a narrow loch, Loch a' Ghille Ghobaich, three-quarters of a mile long, runs by a separate burn into the estuary of the River Morar, on the right bank. Then, still on the north side of the loch, running into the burn from Loch an Nostarie, less than a mile from its foot, we have a tributary from Loch Eireagoraidh, half a mile long. Further east, there is a small burn from Loch na Ba Glaise, with a length of a third of a mile. Lastly, on the north side, a double loch of nearly three-quarters of a mile in length, Lochan Stole and Lochan Ropach.

On the south side of Loch Morar, and running into it, there are three fair-sized lochs, all of which are high – Lochan Chneamh (over 1000 ft., with a small loch running into it), Loch an a' Choire Rhiabhaich (1177 ft., also with a small loch) and Lochan a' Bhrodainn (1234 ft.).

Nan Uamh

We now come to the basin of the sea-loch, Loch nan Uamh, where three freshwater lochs have been measured: Loch Dubh, Loch Mama and Loch na Creige Duibhe. Going along the coast to the head of

Loch nan Uamh, we come to the last two which together enter the sea loch through Gleann Mama. Loch Mama and Loch na Creige Dhuibhe were doubtless, at no distant date, one sheet of water that was gradually separated into two portions by the gradual deposition of material brought down by the Allt Dearga. This is evidenced by the fact that locally the name Mama is applied to both divisions, but here that name is restricted to the western basin, the name na Creige Dhuibhe being applied to the larger and deeper eastern basin. The connecting stream is about 60 yards in length, with a depth of 7 to 8 ft., the fall from Loch na Creige Dhuibhe being less than one foot. The tract of alluvium separating the two lochs is about two and a half feet above the water of Loch Mama, and the keeper stated that he had often seen it flooded when the lochs were high.

Loch Mama is over one third of a mile in length. It lies at a height of 359 ft., with a maximum depth of 44 ft. and a mean depth of 14¼ ft. Loch na Creige Dhuibhe is four-fifths of a mile in length. It was at a height of 359.7 ft. on the date of the survey (July 11, 1902), and has a maximum depth of 93 ft., a mean depth of 32½ ft., and a drainage area of 1½ square miles. 'It is much deeper than Loch Mama and the deeper water approaches nearer to the west end than to the east end.'

The third loch that has been measured is peculiar and rather interesting. It is thus described in the Bathymetric Survey:-

> Loch Dubh is a small loch situated at the head of the peninsula of Ardnish, which separates Loch nan Uamh from Loch Aylort, the two branches of the Sound of Arisaig. The Mallaig extension of the West Highland Railway runs along its southern shore, and the outfall flows through the old bed of the little Lochan Deabhta, which has been completely drained by the railway leaving only a channel through for the escape of the waters from Loch Dubh. After leaving Lochan Deabhta, the outfall joins the Schoolhouse Burn, which has been deflected thence into the Arnabol Burn, falling into the head of Loch Beag, an inlet of Loch nan Uamh.
>
> It is surrounded on the western side by low though steep hills, which impart a dark and sullen appearance to the loch, hence the name Loch Dubh – the Black Loch. Considering its superficial area, it is the deepest loch visited by the Lake Survey. Its catchment area is very small, and it would seem that the unpleasant taste of the water, resembling that of a stagnant pool, is due to the small amount of fresh water entering it. This unpleasantness is probably something more than mere taste, for attempts to stock the loch with trout have been unsuccessful, the fish rapidly dying; eels, however, abound in it.

The height of Loch Dubh is 103 ft., while the length of it is under half a mile, the maximum breadth being one-sixth of a mile. Its waters cover an area of 32 acres, while it drains an area eight times greater or about 262 acres, equal to a little more than a third of a square mile. The maximum depth observed was 153 ft., which bears the ratio to the length of the loch of 1 to 15. This low ratio is only equalled by the little loch on Eilean Subhainn, an island in Loch Maree, and the loch which most nearly approaches it is Loch Fender in the Tay basin in which the ratio is 1 to 22, followed by Loch Dhugaill, near Kishorn, in which the ratio is 1 to 27. Among the larger Scottish lochs, the nearest approach is found in Loch Treig with a ratio of depth to length of 1 to 62. The mean depth of Loch Dubh is 62 ft. or 41 per cent of the maximum depth, which was observed towards the north-eastern shore, giving a slope of 59°. The height of the hill immediately adjoining is 240 ft., and the slope 35°; hence the slope from the top of the hill to the bottom of the loch is one of 45°.

The temperatures at 3 pm on July 2, 1902, were:

Depth	°F.
Surface	59.0°
16 ft.	68.9°
20 ft.	56.0°
25 ft.	53.7°
35 ft.	51.5°

50 ft.	47.1°
75 ft.	44.1°
100 ft.	43.6°
150 ft.	43.5°

In March 1903, when the loch was revisited, it was found to be uniform in temperature from surface to bottom.

> Compared with the temperatures recorded in Loch Shiel a week earlier in the same month we find the temperatures in Loch Dubh 1.7° lower at the bottom in 150 ft. than in Loch Shiel in 420 ft. and in Loch Morar ten days earlier in the month. A temperature as low as that at the bottom of Loch Dubh was recorded only after descending to a depth of 250 ft. The extraordinary temperature conditions observed in Loch Dubh may probably be accounted for by (1) the great depth of the loch compared with other lochs of similar area, (2) the small extent of its drainage area, so that very little rain water enters the loch, and (3) the small area of the loch and the steepness of the surrounding hills reducing the mixing effect of the wind. (B.S.).

Before leaving the point of Ardnish, there is another considerable loch that we must notice, which the B.S. had to omit because there was no boat on it. Loch Doir' a' Ghearrain, whose outflow is by a separate burn at the end of the point. It is very nearly a mile long and has a small loch running into it.

Ailort

In the next basin, that of the River Ailort, only one loch needs to be considered, and it alone has been measured by the B.S. - the other lochs in the basin being small and negligible.

> Loch Eilt was formerly considered a good loch for salmon and sea trout, but Mr Harvie-Brown believes that the blasting operations during the construction of the Mallaig extension of the railway resulted in the destruction of a large amount of spawn and fry, and that now the fish are greatly disturbed by the passage of the trains across the bays on the south shore. (B.S.).

The loch is 3⅓ miles in length, at a height of 96 ft. and has a maximum depth of 119 ft., the mean depth being 37 ft. It is naturally divided into three portions by two narrow constrictions in its outline, the western portion being by far the largest and deepest. The western portion is connected with the central portion by a channel 6 ft. in depth, with a rocky islet in the centre, the sides of the channel being also of rock in situ, thus dividing the loch into two rock basins. The central portion is separated from the eastern portion by detritus brought down by the Allt a' Choire Bhuidhe, the channel between them having a depth of 7 ft. The small eastern and central basins are quite simple in conformation, the maximum depth observed in the eastern one being 52 ft., and in the central one, 70 ft. The 75-foot area is about half a mile long, and the 100-foot area about a quarter of a mile in length, occupying the wide central portion of the western basin, but rather nearer the east than the west end.

The temperatures observed on July 10, 1902, were as follows:

Depth	West basin (5 pm) °F.	Central basin (3.30 pm) °F.	East basin (2.30 pm) °F.
Surface	60.0°	58.1°	58.0°
35 ft.	59.7°	53.6°	52.8°
50 ft.	54.8°	52.8°	50.7°
75 ft.	53.0°	-	-
100 ft.	51.0°	-	-

Besides ordinary trout, the loch contains large trout (presumably ferox) which run up to 9 lb. at least, and char. The last are said by Mr G.N. Nall to be small, and to be found in the 'upper section' (*Life of the Sea Trout*).

Shiel

Loch Shiel is the fourth longest loch (17½ miles) in Scotland, Loch Awe, Loch Ness and Loch Lomond taking precedence in that respect. The River Shiel runs into the sea at Loch Moidart from Loch Shiel after a course of two miles. As the B.S. points out, 'its elevation above the sea is only 11½ ft. so that a slight subsidence of the strip of land through which the River Shiel flows would convert it into an arm of the sea. Seals occasionally make their way into this loch at the present time.'

The principal upper portion of the loch trends in a north-east and south-west direction, but about six miles above the outflow there is a bend in the outline of the loch, and the lower portion trends almost due west. At the foot of the loch, the surrounding ground is low, but on proceeding up the loch we find mountainous country bordering it on both sides, culminating in heights exceeding 3000 ft. at the head of Glen Finnan. To the south rises Beinn Resipol (2774 ft.) between Loch Shiel and Loch Sunart; to the east rise many hills over 2000 ft. while in the north there are two hills over 3000 ft.

The principal feeders are the River Finnan, the Amhainn Shlatach and the Callop River, which enter the loch at its head, the Glenaladale River entering about six miles down on the north-western shore, and the River Polloch (bearing the outflow from Loch Doilet) entering 11 miles down on the south-eastern shore, where the bend in the trend of the loch occurs. There are numerous small islands and a few larger ones, the largest, Eilean Gleann Fhionainn, at the head of the loch.

The maximum depth recorded in Loch Shiel is 420 ft., occurring about four miles from the head of the loch, between the heights of Beinn Odhar Bheag (2895 ft.) to the north-west and of Meall nan Creag Leac (2475 ft.) to the south-east. The mean depth is 133 ft.

'The floor of Loch Shiel is on the whole rather irregular.' After the irregularities, the Survey generalises and points out that 'in Loch Shiel the deeper water occurs towards the head of the loch. Proceeding from Acharacle at the foot of the loch, one must row two miles before encountering a depth of 50 ft.; a farther 1½ miles before meeting with a depth of 100 ft. and this merely a small patch; a farther 1½ miles having to be traversed before reaching the main 100-foot basin or a total distance of five miles from the foot of the loch. The main 200-foot basin is distant about nine miles, the lower 300-foot basin nearly 10 miles, and the principal 400-foot basin over 12 miles from the foot of the loch.'

Temperatures in the loch about three miles from its head at 5 pm on July 9, 1902, were:

Depth	°F.
Surface	56.5°
50 ft.	56.0°
100 ft.	47.4°
200 ft.	45.2°

The Survey notes that:

> Salmon, grilse, sea trout and brown trout abound in the loch and yield fair sport, some of the salmon and trout being very heavy.

The only considerable tributary loch is Loch Doilet and it alone was sounded. It is situated on the River Polloch, about 1½ miles above its mouth, at a height of 22 ft. above the sea, and is nearly 1½ miles long. The maximum depth is 55 ft., with a mean depth of 23½ ft. The loch contains salmon, sea trout and ordinary trout.

X

ALINE, LOCHY, ARKAIG, SPEAN AND LEVEN

Aline

Next is the basin of the River Aline, which enters the sea at the head of Loch Aline in Morvern, near Ardtornish. Two lochs have been surveyed, both are practically on the same level (36 ft. above the sea). The first is Loch Arienas, about two miles from the foot of the river. It is a fairly large loch about two miles in length and 116 ft. deep. The second loch is Loch Doire, a smaller water and less deep (48 ft.).

Mull

Here we cross over the sound from Morvern to Mull and note the statistics of the two lochs on Mull which are given by the B.S.

The longer loch, Loch Frisa, in the basin of the Aros River, is four and a half miles long and nearly 500 ft. above the sea. Its maximum depth is 205 ft.

The other loch, Loch Ba, is about two miles south of Salen on the east coast of the island, and is at a height of fully 40 ft. above the sea. The loch is three miles in length, and has a maximum depth of 144 ft., and a mean depth of 47½ ft.

The soundings show that the floor of Loch Ba is
somewhat irregular, due principally to the fact that a
shallow ridge crosses the loch at its narrowest part, a
little more than a mile from its southern end. Here,
the breadth is only quarter of a mile, and the deepest
sounding recorded on the ridge was 60 ft. The 25-
foot and the 50-foot areas are thus continuous, and
extend nearly the whole length of the loch, but the
75-foot area is cut into two portions, the smaller
portion to the south-east of the ridge having a
maximum depth of 95 ft. while the larger portion to
the north-west of the ridge includes the deepest
water in the loch, the 100-foot basin being nearly a
mile in length.

Lochy

Next there is the basin of the River Lochy, which
enters the sea at the head of Loch Linnhe. The two
main lochs here are Loch Lochy and Loch Arkaig.

We naturally begin with Loch Lochy, which is
situated on the River Lochy at Gairlochy, about
eight miles from its foot. The loch serves as the
most southerly section of the Caledonian Canal, and
is a little under 10 miles in length, its height above
the sea being 94 ft. It is a very deep loch, the fourth
deepest (531 ft.) in Scotland, with the considerable
mean depth of 229 ft. The hills on the west rise with
a uniform very steep slope to heights of about 3,000
ft. - Ben Tee (2957 ft.), Sron a' Choire Ghairbh

(3066 ft.) and Meall Choire Lochain (2971 ft.) –
broken only by the deep gashes torn by the torrents
in the glacial debris, which here extends far up the
mountains. On the east, the slope is about the same,
but the hills are less high, the ridge (almost wholly
covered with debris) which separates Loch Lochy
from Glen Gloy, reaching to 2,000 ft., where there is
a trace of a third parallel road. The only important
streams feeding the loch are the River Arkaig (from
Loch Arkaig) entering near the lower end of Loch
Lochy and the River Gloy, the rest of the feeders
being mere mountain torrents. A very small part of
the overflow from Loch Oich enters Loch Lochy by
the Caledonian Canal.

At the north end, a small basin called Ceann Loch,
measuring one half by one third of a mile, and
having a maximum depth of 66 ft., is cut off from
the main loch by a narrow channel in which the
greatest depth is 40 ft.

The main loch is a simple basin, with the U-shaped
section characteristic of glacier formed lakes. All the
contours are continuous, those at 50 and 100 ft.
enclosing areas little less than the total length of the
loch. The area enclosed by the 200-foot contour
measures 6½ miles in length, and by the 300-foot
contour a little over three miles in length. From
opposite the mouth of the River Arkaig the loch
shallows rapidly, and the contours are irregular. The
flat-bottomed character of the basin of Loch Lochy

is indicated by the comparatively large area covered by water between 400 and 500 ft. in depth.

Above Loch Lochy is Loch Arkaig at a height of 140 ft. It is a long, narrow, curving loch, which tapers towards each end. Its length is 12 miles and it has a maximum depth of 359 ft., with a mean depth of 153 ft.

The basin of Loch Arkaig is nearly simple, the slight irregularities doubtless correlated with the curving outline. The contours at 50 and 100 ft. are continuous. A little over two miles from the west end of the loch there is an abrupt narrowing, and the loch continues to narrow to the end. Corresponding with this, the 200-foot contour is broken into two basins. In the narrow eastern part is a separate 200-foot basin with a maximum depth of 262 ft., this is only separated from the main 200-foot basin by a slight shallowing to 183 ft. The main 200-foot basin is about eight miles long. Though the wide portion of the loch, fully nine miles in length, forms a simple basin, there is not the well-marked U-section found in typical glacier-formed lakes.

The water temperatures in Loch Arkaig on 11 June, 1902, were:

Depth	°F.
Surface	49.4°
25 ft	47.5°
50 ft.	46.5°
100 ft.	45.5°
150 ft.	45.0°

A few records of the fish in Loch Arkaig are given by the Third Earl of Malmesbury in his *Memoirs of an Ex-minister*, which he published in 1884. From 1844 to 1859 he rented from Lochiel the house of Achnacarry along with the deer-forest at Loch Arkaig. He had been Foreign Secretary in 1852 and again in 1858-59.

Malmesbury records that on:

> Oct. 20, 1853 – Grantley Berkeley arrived [at Achnacarry] and was as agreeable as he always is; but considering his great reputation as a sportsman, he did nothing in deer-stalking, being past the age for walking over Lochiels's mountains.

> Oct. 28 – Berkeley killed a *Salmo ferox* weighing 18 lbs. in Loch Arkaig.

The following passage is interesting:

> Nov. 5 – Went to the forest of Gerraran [Coille nan Geur-oirean], a primaeval wood stretching

along the southern shores of Loch Arkaig, and
killed a magnificent stag with twelve points, a cup
on each horn, and double brow antlers. This wood
and that of Gusach [Coille a' Ghiubhais], lining
the shore of Loch Arkaig are certainly primaeval.
The hill is clothed with immense pines, and with
almost impenetrable heather. Among the debris of
centuries and in an older stratum lie many gigantic
oaks; one I measured was sixty feet long and
perfectly sound. They were evidently the ancient
possessors of the mountain, before the younger
generation of the red pine usurped their place.

The following entry, however, presents a problem:

Sept. 23, 1854 – I went out fishing on Loch
Arkaig and caught a bull trout of 18 lbs. and Lord
Edward Thynne a salmo ferox of 13 lbs.

Does this imply a deliberate intention to draw a
definite distinction between a 'bull trout' and a
'salmo ferox'? It may be only a casual variation of
phrase, but some of the larger fish in Loch Arkaig
are rather ambiguous in appearance. They are gray
in colour and on the whole resemble large sea trout
rather than ferox, the usual characteristic of which is
the copper colour on the sides of the body.

The remaining reference to Malmesbury does not
give us any additional light on the question:

Sept 30, 1859 – Lady Chesterfield and Lady
Evelyn Stanhope arrived. The former caught
several salmo ferox in Loch Arkaig.

except perhaps to emphasise the fact that the name
'bull trout' is used of only one specimen.

The following seems of sufficient interest to quote:

Oct. 3, 1837 – This morning my stalker and his
boy gave me an account of a mysterious creature
which they say exists in Loch Arkaig and which
they call the Lake-horse. It is the same animal of
which one has occasionally read accounts in the
newspapers as having been seen in the Highland
lochs and on the existence of which in Loch
Assynt, the late Lord Ellesmere wrote an
interesting article, but hitherto the story has always
been looked upon as fabulous. I am now, however,
nearly persuaded of its truth. My stalker, John
Stuart, of Achnacarry, has seen it twice, and both
times at sunrise in summer on a bright sunny day,
when there was not a ripple on the water. The
creature was basking on the surface; he only saw
the head and hindquarters, proving that its back
was hollow, which is not the shape of any fish or
of a seal. Its head resembled that of a horse. It was
also seen once by his three little children, who
were all walking together along the beach. It was
then motionless, about thirty yards from the shore
and apparently asleep, and they first took it for a
rock, but when they got near it moved its head,
and they were so frightened that they ran home,

arriving is a state of the greatest terror. There was no mistaking their manner when they related their story and they offered to make an affidavit before a magistrate. The Highlanders are very superstitious about this creature. They are convinced that there is never more than one in existence at the same time, and I believe they think it has something diabolical in its nature, for when I said I wished I could get within shot of it my stalker observed very gravely: 'Perhaps your Lordship's gun would miss fire.' It would be quite possible, but difficult, for a seal to work up the River Lochy into Loch Arkaig.

Five small streams enter Loch Arkaig. The River Dessary and the River Pean together enter at the head, the former rising near the conspicuous Sgurr nan Coireachan (3125 ft.), the latter in the small Lochan Leum an t'Sagairt near a slightly higher hill (3136 ft.) of the same name. Then, two and a half miles down the loch on the south side, the Allt Camgharaidh comes down from the small Lochan a' Chomhlain, near the highest hill in the neighbourhood, Culvain (3224 ft.); the hill on which the Earl of Dalkeith was killed. Farther down on the north side, the Allt Arcabhi comes down steeply from a trout loch, Loch Blair, which is nearly three-quarters of a mile long (at a height of 993 ft.) and, moreover has two small lochs running into it. Lastly, nearly three miles from the foot, there enters the River Mallie.

The B.S. has some interesting notes on the loch.

> The lower part of the loch is well wooded,
> picturesque and romantic, with hills to north and
> south reaching well over 2,000 ft. in height. The
> upper part is barer and grander, the mountains
> exceeding 3,000 ft. in height. A road runs along
> the north side of the loch, deteriorating towards
> the west end into a rough track which leads to
> Loch Nevis and Loch Morar. Several wooded
> islands enhance the charm of the scenery, and on
> one of these is one of the few nesting places of the
> osprey, still occupied by the birds at the time that
> the survey was made.

There is very good fishing in Loch Arkaig, and lake
trout up to 10 lbs. in weight were taken while the
survey was going on.

It may be added that there are pike in both Loch
Lochy and Loch Arkaig, as well as salmon, sea
trout, ferox and ordinary trout, but no record of char.
The canal steamers use Loch Lochy, and a private
steamer was put on Loch Arkaig before the end of
the last century.

Spean

The River Spean joins the River Lochy near
Gairlochy and has several large lochs within its
basin. However, these lochs have been covered in

Part I of *Ferox and Char*. They include Loch Treig, Loch Ghuilbinn and Loch Ossian.

Leven

Four lochs in this basin were surveyed in May, 1903, but shortly afterwards three of these lochs were formed into a reservoir for the supply of water to an aluminium factory at Kinlochleven, a few miles down the river.

These three lochs were in sequence at heights of 992, 1022 and 1024 ft. The lowest was the deepest (50 ft.).

The fourth loch surveyed, Loch-Eilde Mor is on a tributary from the north at a height of 110 ft., and is about two miles long. The maximum depth is 100 ft., with a mean depth of 47 ft. The loch is near Binnein Mor (3700 ft.) and Sgurr Eilde Mor (3279 ft.). The loch was surveyed on May 13, 1903, when the surface temperature was 47°F. It has a somewhat flat-bottomed character and is used to some extent to supplement the main reservoir.

The S.G. says that trout have been killed in the three lochs on the Leven, the Black Lochs, as they were called, up to heavy weights, and that both Loch-Eilde Mor and the smaller Loch-Eilde Beag, which are in the Forest of Mamore, yield good sport. The latter is half a mile long and runs into the former.

XI

ETIVE AND AWE

Etive

The B.S. surveyed other Black Lochs (three, or rather four, expansions of the Lusragan Burn), stretching for about two miles in a north-east direction near Connel Ferry. These have basins which decrease in depth as we proceed up the burn. A maximum depth of 36 ft. was observed close to the north end.

The remaining four lochs in the Etive basin, enumerated by the lake survey (one being a group of four), seem too unimportant to be worth naming. Two of them are on the north side of Loch Etive.

Awe

The B.S. describes Loch Awe as

> extremely elongate, but sinuous in outline, and is peculiar in that a long narrow arm branches off at right angles to the main axis, and leads through the Pass of Brander to the outflow. Loch Awe exceeds in length all other Scottish lochs, for measured along the central axis from the head of the loch to the exit of the loch in the Pass of Brander, it is almost 25½ miles in length.

Its height above the sea is 118 ft., the maximum depth being 317 ft., and the mean depth about 105 ft. But originally it was deeper and larger. At some time in the past, the erosive activity of the River Brander succeeded in 'tapping' the loch and lowering it by more than 100 ft., thus shortening the loch by several miles, especially near its original exit into the River Add and Loch Crinan.

The floor of Loch Awe is uneven. The 50-foot contour line is continuous, and coincides, on the whole, with the outline of the loch, extending nearly from end to end. The 100-foot contour encloses three separate basins. The largest basin extends for little more than a mile from the head of the loch to near the islands at the junction of the arm at the Pass of Brander, a distance of about 18 miles. The second basin extends from about half a mile from the entrance of the River Orchy into the arm at the Pass of Brander and is over four miles in length. The third, and smallest basin, based on two soundings in the Pass of Brander, is separated from the second basin by a depth of 86 ft.

The 200-foot contour encloses no fewer than five separate basins. The principal area, including the deepest water in the loch, about 5½ m. in length, extends from the entrance of the Kames River to little more than two miles from the head of the loch. It is to be noted that the middle portion of the loch,

between Portinnisherrich and Ingstrinich – a stretch
of over twelve miles, is less than 200 ft. in depth.

The 300-foot contour encloses an area about 1½ m.
in length, less than five miles from the head of the
loch. Over 57 per cent of the lake floor is covered by
less than 100 ft. of water.

There are early references to fishing in Loch Awe in
Stoddart's first book, which was called *The Art of
Angling, as Practised in Scotland*, and was
published in 1835, when he was twenty five years
old.

> We now come to speak upon the different species
> of freshwater trout, and primarily upon the great
> lake trout or salmo ferox of the naturalist.

At the time, as we have seen, it was generally
believed that the different kinds of trout were
distinct species of the fish, and not merely different
varieties of it. This view Stoddart held at first, but
he afterwards tended to modify it.

> This fish is erroneously supposed by some to be
> confined, in Scotland, entirely to Loch Awe. But it
> exists in very many highland lochs. It is found in
> Lochs Laggan, Ericht and Garry. It has been
> caught, over and over again, in Rannoch, Tummel,
> Aich and Lydoch.

In the second edition of the book (1836), this list of lochs that contain ferox is somewhat altered. Two are omitted, Lochs Ericht and Aich (= Eigheach on the Moor of Rannoch); the omission of the latter is perhaps defensible as it is too shallow for the great lake trout, but not that of the former. Three are added: Lochs Quoich, Monar and Shin, the addition of Loch Monar being almost certainly wrong.

> To a Mr Morrison, from Glasgow, is attributed the merit of having first discovered the salmo ferox in Loch Awe, about half a century ago. We doubt very much the strengths of his claims to this discovery; and from inquiries made by us at Dalmally, Cladich, Inverawe and other parts of the surrounding country, we are led to believe that this species of trout has been known there from time immemorial.

> This fish acquires prodigious dimensions. One was caught in Loch Rannoch, by the late Baron Norton, weighing thirty pounds. In Loch Awe they have frequently been taken by Mr Maule of Edinburgh, and others, betwixt twenty and twenty-eight pounds; while on the continent in Norway and Switzerland it is not uncommon to meet with them weighing nearly four stone (56 lb.). In America, trout have been captured of a still more incredible size. The salmo ferox, like the pike, is a strong, fierce, fish and when attaining a weight of two pounds begins to despise your flies, and becomes a sort of cannibal, preying upon its own species with much rapacity. Its proportions, when

large, are somewhat singular, the depth of the fish being astonishingly great when compared with the length. A thirty pound lake trout seldom measures a single yard, while a salmon of equal weight is considerably larger.

We may compare Stoddart's ideas about what he regards as the primary type of trout with those of a modern authority, such as W.J.M. Menzies in his *Sea Trout and Trout* (1936):-

> From the Traun See (8 miles long and 1385 ft. deep) of the Austrian Alps have come trout of 45 lbs., and in this present year (1935) others of 47 and 42 lbs. From the Lunzer See another of 44 lbs. was taken in 1931 and from Lake Lugano and Vilalpsee two of 39 lbs. each, one in 1921 and the other in 1927 respectively.

The British records are given as follows:

> The only others able to compare at all with these Goliaths of the trout family are from a few Irish loughs. Well authenticated records exist of trout of 30½ lbs. from Lough Derg and 26 lbs. from Lough Ennell in the Free State. In Northern Ireland where Lough Neagh is the home of similar trout, which, however, are here taken with net (unlike those mentioned in the Free State) and usually about August when they come out of the deeper water to enter spawning tributaries. I have been told of one of 34 lbs., and apparently a certain number of 20 lbs. and 30 lbs. are occasionally taken.

Real trout of 22 lbs. in 1867 and 21 lbs. in 1904 have, however, been taken in Loch Rannoch, and another good fish from the same loch in 1912 weighed 18½ lbs. Quite a number of trout from this last size downwards also have been taken here. Loch Garry (Ness basin) and the adjacent Loch Poulary form another habitat of the true brown trout weighing up to 15 lbs. Loch Awe is yet another home of these exceptional fish (e.g. 16 lbs. in 1935).

From time to time, trout between 10 and 20 lbs. are also take in two quite small lochs close to the village of Durness in the far north-west corner of Scotland. The lochs are situated on the limestone ridge, which extends from that district in a south-westerly direction and this no doubt accounts for the size of the trout.

The estuarine trout enter into this competition of giants. Among these, the specimen of 29 lbs. taken on a plebeian long line in the Loch of Stenness in Orkney in 1889 holds pride of place. In that curious tidal loch with a mixed population of trout, it may have really been a true sea trout, although described at the time as a 'slob' trout.

Stoddart concludes his brief remarks on the salmo ferox by giving the record 'bag' of these fish:

The greatest number of these fish known to have been captured by the rod in one day was thirteen. They were taken in Loch Awe as early as March of the beginning of April some years ago by a Mr Lavrock of Keswick, and weighed altogether above ninety pounds.

It is interesting to compare this 'bag' with that of Sir Alexander Gordon Cumming in the Fionn Loch near Gairloch in 1851. Each consisted of about the same number of fish, and both were taken early in the year. The more southerly catch is two or three pounds heavier than the other, and probably includes a larger ferox – perhaps 15½ lbs. as against 15 lbs.

A friend of Stoddart, James Wilson, the author of *The Rod and the Gun*, tells us that:-

> This huge species may be said, indeed, to be by far the most powerful of our freshwater fishes, exceeding the salmon in actual strength, though not in activity. The most general size caught by trolling, ranges from 3 to 15 pounds, beyond which weight they are of uncommon occurrence. The largest recorded to have been killed in Loch Awe amounted to 25 pounds, and the heaviest we have lately heard of as captured there, was a few ounces under 20 pounds.

He says, incorrectly, that even 'when in perfect season the head is too large and prolonged to be in accordance with our ideas of perfect symmetry in a trout.' But he is more successful in trying to describe the colour and points out that 'the whole body, when the fish is newly caught, appears as if glazed over with a thin tint of rich lake-colour.' though perhaps one would rather be inclined to speak of a bright copper tint like that of a new penny.

I will supplement these statements by Stoddart and Wilson about Loch Awe by quoting John Colquhoun, the well-known author of *The Moor and the Loch*. The first edition was published in 1840, the third in 1851 and the fourth in 1878. In the preface to the fourth edition, the author draws a contrast between the circumstances in which it was written and those of the previous edition:-

> Perhaps a more favourable observatory of this kind could scarcely be found than my present fascinating home on the banks of Loch Awe. The crow of the moorcock is heard from our windows – the bell of the roebuck, in the adjacent hanging wood, sounds close to our door, a good eye and glass may command the corries of Ben Cruachan; infinite varieties of wild fowl crowd our loch in winter, many of them rare polar visitors; the salmon streams of the Awe and Orchy are within easy distance, and the mighty Ferox roams our shores for miles.

It is from the preface of the fourth edition that I wish to quote Colquhoun's description of a day's trolling in Loch Awe in 1845 long before, about 1878, he had come to reside in Edinburgh and was able to make only occasion 'raids' into the Highlands:-

I had already bespoken the services of old Sandy MacKenzie, 'wha kens whar the big fish lie as weel as ony man on Loch Ow side'. Sandy being appointed skipper, begged to be allowed to choose his own crew, which consisted of a stout, good natured 'callant' of about sixteen yclept 'Johnny', occasionally 'Jock', when Sandy was in a patronising mood. Sandy was once a strong bony man, and piqued himself on being one of the best wrestlers in the country. Now his eye is dim and filmy, and his athletic arm is paralytic and weak as a child's. I might have had a far abler man at the oar and as knowing about the habits of the fish; but whenever I troll Loch Awe, none but that poor, ragged, woe-begone old man shall command my boat so long as he is able to do it.

Having satisfied myself that the cobbles were not more leaky than they generally are, I returned to the inn... All was now ready for the evening fishing. Johnny carried my trolling rods; Sandy a 'cogue fu' of live bait and a little basket of provisions; and I my duck-gun.

The rods were soon baited, the evening was perfect for fishing, and the feroxes took well. We came over no large ones, however, and the three brought into the boat were only four, three and two and a half pounds. We had intended to troll to the bay of the old Castle Connall, eight miles down the loch, built, as Sandy says, by the Danes, but

were obliged to defer it to the next day. The bay
which this castle commands is a famous resort of
the largest size of the ferox.

The night overtook us before we could gain the
harbour of Cladich, and the old dun eight-day
clock had just chapped seven, when my gallant
crew cleared out of harbour and, with my rods,
bait provisions, and pea-jacket, were making for
Port Sonachan quay, where I had directed them to
meet me. The morning was colder, the wind had
changed from west to east – a 'bad airt' for the
fish. There were certain appearances also in the
sky which foreboded squally weather. The best of
the fishing ground is below Port Sonachan, so I
did not wish to waste it on such an unpropitious
day until we got there. I sauntered dreamily along,
admiring the views as they unfolded themselves,
listening to a chorus of cuckoos, before the
measured stroke of Sandy and Johnny appeared at
some distance, slowly propelling their clumsy
boat. I was soon seated at the stern, with lovely
baits towing behind me – 'no a rug', as Sandy
repeatedly said, but he endeavoured, poor fellow,
to keep up our spirits by telling a tale of every
wood, or rock, we crept slowly past.

My reveries were now broken by Sandy pointing
out the nest of the 'salmon-tailed gled' and there
are the owners wheeling their graceful circles.
Two roes were also looking at us from the shore,
and another a little further on. They seemed not
the least afraid as we pulled slowly past. I was
admiring the beautiful hanging wood in which the

kite's nest held a prominent place near the top of
one of the finest old oaks, when a pull, that bent
my rod's top to the water, and spun round my
large wooden pirn brought me to my legs at a
spring. To seize the rod and place the butt above
my knee, with a good bend at the top, was the
work of an instant. Sandy was also active: he gave
both oars to Johnny, and began, with his shaky
hands, to wind up the other rod in case of a
collision. I told him always to do so when I
hooked a trout. At this moment the gorgeous fish
sprang a yard out of the water, coming down with
a splash that made the rocks echo. Sandy, at no
time very expert, became quite nervous at the sight
of the monster, and bungled his work sadly. I gave
him a push out of my way, and in so doing
knocked off his tattered hat into the water at the
bottom of the cobble. He only smiled without a
vestige of anger. I saw his thin grey hair and am
happy to recollect that at that exciting moment,
ashamed of my impatience, I picked up his hat
with my left hand, and placed it on his head, poor
Sandy all the time begging me 'never to heed it'.
Sandy's whole heart was in the capture of the fish.
His rod was by this time wound up, he was again
at his oar, and I had fair play. The ferox bored like
a harpooned whale; sometimes he would change
his course, and go down to the bottom, taking
forty yards of line, which he made swirl through
the water, with a humming noise, like a low sound
of the telegraph wires. When I shook him up, he
would fight away for the middle of the loch. At
length he grew weaker and I got him under control
of a short line. It was a beautiful sight - that noble

fish, showing his glancing scales for a moment
and then trying to bore under the boat, and always
foiled by the boatmen, who promptly obeyed my
slightest signal. He now began really to fail, and I
felt I could lead him; so directing Sandy to a
shingly part of the shore, where there were no
rocks, I determined to land him there. The beach
was very shallow and, in spite of my
remonstrances, Sandy walked up to his knees in
water and drew the cobble ashore. I was now on
Terra firma, but my fish was by no means done up
yet. Every time I brought him to the shallow, he
dashed away with as much vigour as before. This
could not last, and the bursts became shorter and
slower, till my victim was unable to get down at
all, and only struggled on the top of the water. I
had ample opportunity to admire his dimensions,
colour, and shape, and was resolved that no
rashness or eagerness to obtain it should rob me of
so rich a prize. At last he turned up on his broad
and gleaming side. Now was my time. And like a
wrecked and gallant vessel he lay stranded on the
beach.

A proud man was Sandy McKenzie then. He took
entire possession of the fish and would hardly let
Johnny look at it; if he ventured to touch it he met
with a stern rebuke. Well did Sandy know how
rare it was to come across a trout of that size in
Loch Awe nowadays. The gale had freshened. The
squalls had settled into a steady gale, and we were
fully seven miles from Cladich. I wished at least to
try on our way home, having little hope of pulling
the whole distance against such a head-wind.

Sandy however was unable to make any way. I
had relieved the old man when we had to cross the
loch or go quickly past bad fishing ground, upon
which occasions I used to hear Johnny taunting
him. When I took the oar, Sandy always has his
revenge by 'You've met your match noo lad.'

Arrived at Cladich, my first step was to order in
the steel yard, when my fish proved 15 ½ lb. odd,
so must have been near sixteen when taken out of
the water. I had killed in Loch Vennachar the year
before, with single gut, a clean salmon which
weighed 17 lb. when brought home. This salmon
did not make nearly so fierce a run as the Loch
Awe trout with gimp. I have heard gentlemen
speak slightingly of the best trout when compared
with salmon; but let them have one of these Loch
Awe monsters on their hooks, in as good condition
as mine was, and I venture to say that they will not
complain of the want of mettle in the trout. I have
no doubt the *Salmo ferox* is superior, both in
strength and spirit to the *Salmo salar*. Unless the
ferox is in first-rate condition, his head is very
ugly, and looks much too large for his body. This
was not the case with the specimen I have just
described; his head is smaller and his shoulder
more round than any I have ever taken; on which
account I had him preserved by Fenton in
Edinburgh.

The char in Loch Awe are said 'to frequent the head
of the loch, around the place of the original outlet'
(Stoddart, Ordnance Gazetteer of Scotland, 1886).

This agrees with the fact that the deepest water lies towards the head.

One steamer at least was put on the loch before the end of the last century. That presumably brought nearer the end of trolling for ferox, which, as we have seen, was beginning to decline as early as 1840.

Besides Loch Awe, several other lochs in the basin were included in the survey. The largest of these is Loch Avich, which lies little more than a mile to the east of the central part of Loch Awe, and is drained by the River Avich, two miles below the outflow of the Kames River on the east shore and near Innis Chonnell. It is 311 ft. above the sea, and nearly 200 ft. above Loch Awe, the maximum depth being 188 ft. (at two places near the east end) and the mean depth 98½ ft., more than half the maximum.

The conformation of the loch is simple. The 54-foot basin is about 3 miles long, the 100-foot basin about 2½ miles, and the 150-foot basin nearly 2 miles in length, in each case approaching closer to the east than to the west end. The off-shore slope is very steep in places, for instance off the northern shore, about three-quarters of a mile from the east end; a sounding in 82 ft. was taken at 80 ft. off-shore, giving a gradient exceeding 1 in 1. The flat-bottomed nature of the loch basin, with the U-

section characteristic of glacier-eroded loch basins,
is well marked.

Temperatures at 6 p.m. on May 27, 1903, were:

Depth	°F.
Surface	55.9°
10 ft.	50.5°
15 ft.	49.5°
25 ft.	47.0°
50 ft.	45.1°
100 ft.	44.8°

Loch Avich is not accessible to salmon. There
seems to be no record of either ferox or char, though
the loch is suitable for them. Anglers are conveyed
from Port Sonachan to within two miles of the loch
by a motor launch.

Next (in the Loch Etive basin) we may take the
small Loch Ederline, near the head of Loch Awe and
only 4 ft. above it, which must therefore have been
submerged in the original Loch Awe. Loch Ederline
is nearly two-thirds of a mile in length, and has a
maximum depth of 58 ft., with a mean depth of 23
ft. It is said to contain pike chiefly.

Passing to the foot of Loch Awe, on the River
Orchy, which enters at the north-east corner, we find
two lochs. The larger, Loch Tulla, at a height of 542

ft., is 2½ miles in length, and has a maximum depth of 84 ft., with a mean depth of 38 ft. It holds pike, which were introduced into it by one of the Earls of Breadalbane, and some good trout.

The upper loch on the River Orchy, Loch Dochard, three miles up, at a height of 735 ft., is about two thirds of a mile in length and has a maximum depth of 42 ft. It is situated in the Black Mount deer forest, near Beinn Suidhe (2215 ft.), and has two small lochs tributary to it.

Two lochs of no great interest on the west side are directly tributary to Loch Awe and of some size, Loch an Leoid and Loch an Droighinn. The former drains into the latter and thence into Loch Awe by the Kilchrenan Burn. It is at a height of 602 ft., the other loch being a foot or so lower. Its maximum depth is 84 ft., the mean depth being nearly 36 ft. The lower, Loch an Droighinn, has a maximum depth of 42 ft., and a mean depth of nearly 15 ft.

Along with these two lochs we may group Loch Tromlee, as it descends by the same burn. It is given by the S.G. but not by the survey. It is said to have been invaded by pike.

With these lochs we may take the larger Loch Nant, which finds its way into Loch Etive by a separate stream, the River Nant. It is nearly a mile long and is about 606 ft. above the sea. The maximum depth is 92 ft. and the mean depth 24 ft.

The name Sior Loch is given to three shallow little lochs which drain into Loch Nant, and are rapidly becoming bog. They are at a height of 733 ft.

Another smaller loch, Lochan Iasgaich, drains into Loch Nant. Other smaller lochs which are in the Etive basin are given in the lake survey, but they are not interesting.

XII

FEOCHAN, SEIL AND MELFORT

Feochan

Two rivers drain into the sea loch, Loch Feochan – the River Nell into the head and the River Euchar near the foot.

On the former, one loch, Loch Nell, has been measured. It is nearly two miles in length, and at a height of only 49 ft. It may at one time have formed part of the sea loch, being separated from it by low, flat, alluvial ground. The north-eastern half of the loch is comparatively shallow, but the south-western part is deep, the maximum depth of 115 ft. having been recorded little more than half a mile from the lower end. The mean depth is 37 ft.

Stoddart in *The Angler's Companion to the Rivers and Lochs of Scotland* says:

> Loch Nell communicates with the sea by means of a small river called the Clugh. The trout here are very large, frequently six and eight pounds weight. There are besides, numbers of charr.

On the River Euchar, two lochs have been measured. The lower, Loch Scamadale, 2½ miles up, at a height of 221 ft., is the larger, over a mile and a half in length. The maximum depth is 145 ft. and the mean depth nearly 70 ft.

The basin has a somewhat flat-bottomed character, the zone covered by water between 50 and 100 ft. in depth being larger than the shore zone covered by less than 50 ft. of water.

Stoddart says that:

> Loch Scammadale contains more salmon than Loch Nell and plenty of sea-trout. Yellow trout have been caught five or six pounds weight, but average much less.

The higher loch, Loch na Speinge (or 'String') is at a height of 778 ft. and about a mile north of Loch Avich in the Etive basin. It is subtriangular in outline with an apex pointing in a south-west direction, and a large island occupying a central position in the loch. It is half a mile in length, and has a maximum depth of 43 ft. and a mean depth of 17½ ft. The loch is named from an old road, the String of Lorne. The S.G. says it has the reputation of being the best trout loch in the neighbourhood.

The temperatures in Loch Scamadale on June 1, 1902, were as follows:

Depth	°F.
Surface	55.0°
20 ft.	52.1°
35 ft.	47.8°
70 ft.	46.4°
100 ft.	46.2°

Seil

We come next to the Seil basin, where one loch has been measured.

Loch Seil lies to the south of Loch Feochan. It is two-thirds of a mile in length, and is 55 ft. above the sea. The maximum depth is 91 ft. and the mean depth is 37 ft.

The temperatures on June 3, 1903, were

Depth	°F.
Surface	59.0°
15 ft.	54.1°
40 ft.	50.2°
80 ft.	48.0°

Melfort

Next is the Loch Melfort basin, where eight lochs were measured by the Survey.

The largest, Loch Tralaig, is about three miles north-east of Kilmelfort and little more than two miles north-west of Loch Avich. It exceeds a mile in length and is at a height of 420 ft. above the sea. It has a maximum depth of 117 ft. and a mean depth of 41 ft.

Temperatures on June 6, 1903, were:

Depth	°F.
Surface	60.0°
20 ft.	55.7°
45 ft.	47.8°
90 ft.	47.0°

Loch Dubh-mor is less than a mile to the south-east of Loch Tralaig and little more than a mile to the north-west of Loch Avich. It is deep for its size, whose greatest diameter is about a third of a mile. Its height above the sea is about 900 ft. The maximum depth is 114 ft., the mean depth being about 51 ft.

The temperatures on June 5, 1903 are interesting.

Depth	°F.
Surface	57.4°
20 ft.	49.2°
59 ft.	46.0°
100 ft.	44.7°

Loch a' Phearsain is close to Kilmelfort and is nearly half a mile in length, at a height of 226 ft. The maximum depth is 53 ft. and the mean depth 19½ ft.

The S.G. says it has trout running up to 8 lb.

Loch an Losgainn Mor is about a mile south-east of the preceding and 508 ft. above the sea. The length of the loch is over half a mile and it has a maximum depth of 51 ft. The S.G. says it has trout up to 5 lb.

Upper Kilchoan Loch is about half a mile from the north shore of Loch Melfort, a small loch about one third of a mile in length, with a maximum depth of 70 ft., and a mean depth of 29½ ft. It is at a height of 378 ft. above the sea.

The Survey has measured two other lochs in the basin, but they seem to be unimportant.

XIII

BENBECULA, UIST AND LEWIS

Benbecula

No char are reported in Benbecula. Three lochs were surveyed by the B.S. – Loch Langavat, Loch Olavat and Loch Heouravay. All three are, of course, near the sea and less than 50 ft in depth. Loch Heouravay is the deepest (41 ft.), Loch Langavat is 34 ft. deep and Loch Olavat is the shallowest, only 12 ft. deep. The whole island is low lying, Rueval, the highest and only hill being no more than 408 ft. above the sea.

North Uist

There are two or three parts of the mainland of Scotland which may be described as 'riddled' with lochs, such as the region near Rhiconich in Sutherland or the point of Stoer, north of Lochinver. But this kind of country is best exemplified by the island or group of islands called North Uist in the Outer Hebrides, which are the most extraordinary welter or maze of islands and lochs in the whole of Scotland, described by someone as the place 'where the sea is all islands and the islands are all lochs'.

The largest loch in North Uist, Loch Scadavay, may be used as an epitome of the island as a whole.

There is probably no other loch in Britain which approaches Loch Scadavay in irregularity and complexity of outline. It is an extraordinary labyrinth of channels, bays, promontories, and islands. Though it measures 4¼ miles in length, and about two miles in greatest breadth, there is really no broad open water in the whole loch. The ratio of circumference to length will illustrate how very irregular is the form – though only a little over four miles in length, a rough measurement indicates a shoreline length of 50 miles.

The main road round the island now cuts this loch into two parts, which are connected by such a small channel under the road that in time of flood the south (upper) loch may temporarily rise some feet higher than the other, though normally they are at the same level. There is nowhere any considerable depth, the deepest parts occurring as little holes, while the narrows are usually shallow and a lowering of the surface by no more than 6 ft. would divide the loch into a dozen small lochs and a host of little ponds, while a rise of the same amount would vastly increase its area by including all the higher lochs in the same basin, among these such large lochs as Loch nan Eun, Loch Huna, Loch Obisary and Loch Deoravat.

One other loch in North Uist has a greater volume than Loch Scadavay, though of less superficial area, Loch Obisary having twice the volume.

There is a chain of lochs running down into the sea from Loch Scadavay. The first is Loch na Garbh-Abhainn Ard, the second is Loch Garbh Abhainn, the third is Loch Skealtar, the fourth Loch nan Geireann which is tidal. During flood, the second is one loch with the first, but at other times there is a strong current through the narrow part of the upper section (Garbh-Abhainn). The outflow of Loch nan Geireann is a drain under the road into the small Loch na Cista, a sea loch. The only salmon in the island are got in these five lochs which drain from Loch Scadavay.

The first thing we have to do is to try to put down the different basins in which the measured lochs are distributed. This we find on page 203 of the B.S. II ii:

> Drainage areas –
>> nan Geireann (tidal)
>>> Deoravat
>>> Tairbert Stuadhaich
>>> a' Bhuird
>>> Huna
>>> na Moracha
>>> nan Eun
>>> Scadavay
>>> na Garbh Amhuinn
>>> na Garbh Amhuinn Ard
>>> Skealtar

Strumore
  Fada
na Creige Leithe
  Garbh Clachan
Grogavat
  a' Ghlinne Dorcha
Oban nam Fiadh
  Caravat
  an Iasgaich
  'ic Colla
  an t'Seisgain
  na Leitir-Eileana
  na Coinnich
Hunder
  na Chonnachair

As we have just seen, the first or Geireann basin contains salmon and is the only basin in North Uist that has that fish, the chief salmon lochs being Loch nan Geireann, Loch Skealtar and the small tidal Loch Ciste. The two Garbh Abhainn lochs are only expansions in the small river leading from Loch Scadavay to Loch Skealtar.

More interesting is the next basin, that which contains Loch Fada, which has been known for long to contain char. Loch Fada is about three miles from Lochmaddy, and is in the middle of the sanctuary for deer on the island, which are not numerous

Loch Fada consists of two portions connected by a narrow channel. The north portion is narrow and elongate from west to east. It is studded with islands on which large numbers of gulls and other birds nest. The greater part of the loch is less than 10 ft. in depth. The maximum depth of 26 ft. is found close to an island at the east end. There is also a very narrow channel with depths up to 25 ft., between the peninsula called Ard Fhada and a chain of small islands.

The southern portion of the loch is triangular. It is one of the largest bodies of open water in North Uist, but even here there is a heap of stones projecting above the surface almost in the centre of the triangle. There are two holes 45 ft. deep. The shore and the islands are entirely of rock, except at the east end near the outflow where there are some mounds of gravel. The length is fully 1⅔ miles and the mean depth is 10 ft. The drainage area is very small, with only small local burns.

The outflow is by a very small stream into Loch Galtarsay whence a river half a mile in length leads into Loch an Strumore which is tidal. The height of the loch is 30 ft.

There are many trout in Loch Fada, and also char. Tate Regan thinks that the char of Loch Fada is closely allied with *S. willughbii*. The ordinary trout in Loch Fada average nearly half a pound. At the head of the loch is a small tributary loch, which is said to have large trout. Loch Hungavat, a considerable loch, also runs into Loch Fada from the north.

Loch an Strumore, a tidal loch, runs out of Loch Fada. The bottom is almost level, about 12 ft. deep in the central parts, the bays shallower. It is 3½ ft above the sea.

The next basin has two lochs, Loch na Creige Leithe and Loch nan Garbh Chlachan. The latter is all shallow with a maximum depth of 25 ft., the maximum depth of the former being 114 ft.

The Loch Crogavat basin is more interesting, for the upper loch, Loch a' Ghlinne-Dorcha, is the second deepest loch in the island (95 ft.). It is a small dark loch occupying the whole eastern flank of Burrival (461 ft.); the mean depth, 27½ ft., is the greatest in the island.

We may here, before proceeding to consider the Oban nam Fiadh basin, the most interesting in North Uist, insert a notice of Loch Obisary, which is the largest loch, being twice the size and volume of Loch Scadavay. It is by far the deepest loch, being twice as deep as Loch a' Ghlinne-Dorcha.

Loch Obisary lies at the foot of Eaval, the highest hill (1139 ft.) in North Uist, which it half encircles.

It is roughly crescent-shaped and measures 2⅓ miles
in a straight line between the points of the crescent.
There are large islands in the northern part of the
loch, and the broadest open water is only half a mile
in breadth. In the narrow western part there are three
basins with maximum depths of 51, 57 and 50 ft.
South of Eilean Leithann is a basin with a depth of
65 ft. Between Eilean Mor and the stream flowing
out to the north into Loch Eport lies the deepest
basin in the loch. It is of very limited extent,
measuring only a quarter of a mile each way, but has
the remarkable depth of 151 ft. The mean depth of
25¾ ft. is less than that of Loch a' Ghlinne Dorcha
and a little more than that of Loch Crogavat. The
greater part of the shore is of rock, forming on the
west a range of cliffs. Immediately under the north
slope of Eaval is a large stretch of peat-covered
gravel. The height of the loch is 8 ft. Though the
surface level is little affected by the tides, these
enter often enough to render the water quite salt, and
to permit numerous marine animals to live in it.

The temperatures on June 25, 1904, were as follows:

Depth	°F.
Surface	55.5°
10 ft.	55.5°
25 ft.	50.4°
50 ft.	47.1°
150 ft.	48.2°

I now go on to the Oban nam Fiadh basin, in which
Loch Caravat is much the most important loch. Loch
Oban nam Fiadh itself, the lowest loch, is of unusual
form among the lochs of North Uist. It is elongate
and narrow, over a mile long and is divided into
three portions by narrows. The main part is of
oblong form. There are several small islands, one in
the centre of the loch. The bottom is uniform, 5 or 6
ft. in depth. The middle and upper portions are small
and 4 or 5 ft. in depth. They are separated by a small
island and the whole channel here is grown up with
weeds. The stream from Loch Caravat enters the
upper basin. The shores are of rock.

The loch is interesting from the transition it shows
from salt to fresh water and the corresponding
difference in the fauna and flora of the upper and
lower basins. The lower part is purely tidal. If not
filled by all ordinary tides, it is at any rate so
frequently filled as to allow sea weeds to grow and
marine animals to live. Yet freshwater plants also
grow in this part and mussels are found adhering to
these. The very narrow channel and the dense
growth of *Phragmites* prevent the tides from having
much effect on the upper portion in summer. Here
the water tastes almost fresh and such freshwater
Crustacea as *Holopedium* are found. Yet high tides
must raise this part considerably, as Loch Caravat,
at the time of our visit 2 ft. higher, is filled through

it. The temperature at the surface was 68° F. and at 6 ft. was 68.8° F. on June 7, 1904.

The other char loch is Loch Caravat, which is the third deepest loch on North Uist. Loch Caravat, which is rather complicated, is thus described by the B.S.:-

> The second deepest among the larger lochs of the island, though the little a' Ghlinne-Dorcha is somewhat deeper. In general form it resembles the letter H, there being two narrow portions running east and west connected by a narrow channel running north and south. In the circumstances it is difficult to define length and breadth; a line drawn from the west and along the connecting arm, to the east end of the southern branch would be about 2 miles in length.

The northern branch, nearly 1½ miles in length, is divided into three parts:

(1) That to the east, adjoining the outflow, is quadrate, measuring about a quarter of a mile each way; it has an even bottom, with a greatest depth of 20 ft.

(2) The middle portion is filled with islands, on one of which, Dun Scor, is a Dun. Another, Eilean Dubh Dun Scor, is connected with a larger island on the east by a long causeway. Among the islands, the north branch is deeper than elsewhere, the greatest

depth being 30 ft. The west portion of this branch is three-quarters of a mile long, and very narrow and shallow, having a greatest depth of 11 ft. It is separated from the central part by a large island, connected with the north shore by a causeway and having the channel on the south full of stones and from 1 to 3 ft. in depth. The burn from Loch an Iasgaich enters the west end of this arm. The narrow passage connecting the northern and southern branches of the loch is shallow in the middle but toward the south it rapidly deepens into one of the basins which form the southern half of the loch.

(3) The southern branch of the loch is shorter than the northern, measuring little more than one mile in length, but it is broader and very much deeper. It contains two distinct basins, the best marked basins in the loch, separated by a strait filled with large islands. The west basin is triangular. Though the island on which is the Dun lies well out from the shore, it does not destroy the simplicity of the basin. The slope of the bottom is steeper on the south side, more gradual on the north. The deepest sounding in this basin is 50 ft. The east basin is smaller but deeper. It is fully half a mile long, by a quarter of a mile broad. The slopes of the bottom are about equal on all sides and the deepest part of the loch (74 ft.) is about the middle of the basin. The narrows between the east and west basins is nearly closed by islands of which Eilean Dubh, one sixth of a mile long, is the largest. Large portions of the shore are

stony but rock is exposed in many places and the principal islands are of rock.

The stream flowing out of the north-east corner into Loch Oban nam Fiadh, a rather well known sea trout loch, the lower part of which is purely tidal. It is one eighth of a mile long, and has a fall of 2 ft. At the south-east corner, the burn from Loch 'ic Colla flows in. The superficial area of Loch Caravat is over half a square mile. The drainage area, which includes Lochs an Iasgaich, 'ic Colla, an t'Seasgain and other smaller lochs, is fully three square miles. By volume of water, Loch Caravat is the third largest loch on the island.

The temperatures on June 11, 1904, were as follows:

Depth	°F.
Surface	58.8°
25 ft.	57.2°
35 ft.	56.8°
40 ft.	52.3°
50 ft.	51.2°
72 ft.	51.2°

When surveyed (June 8, 1904), the level of the loch was 7⅛ ft. If there be no error in this measurement, the tide must sometimes enter the loch, and the local gillies stated that this was so. Nevertheless, the water is fresh enough to be drunk. Through so many narrow and shallow channels it is probable that the tides can have little effect on the salinity of the more distant parts of the loch.

Loch Caravat is of great interest to the archaeologist, for the fort of Dun Ban on one of the islands, about 200 yards from the southern shore of the loch, is described by the Commission on the Ancient Monuments of Scotland (*9th Report on the Outer Hebrides, Skye and the Small Isles*, 1928):-

> as possibly the most interesting of the insulated dwellings in the Western Isles, for while it must be classed with the island duns and is one of a plan common in the neighbourhood, it is a mediaeval structure, built with lime mortar and having features in common with the castles of the mainland, with which the builder was evidently familiar.

The islet is entirely occupied by the structure which measures externally 64 by 71 ft. and is D-shaped in plan, the frontal curve lying towards the east being penetrated by a re-entrant containing the doorway. The outer wall, 9 ft. in greatest height, averages 6½ ft. in thickness and is built of rubble, roughly coursed, with pinnings in places, laid in a heavy lime mortar bed of poor quality, containing sea shells. The entrance is set back 12 ft. within a recess similar to the boat harbour noted in other duns. The doorway is 4 ft. wide and the sill is 3 ft. above the foundation and 3½ ft. above the water; the jambs are checked on either side of the door and the bar hole 8 inches square is found in the southern jamb, the corresponding socket being ruinous.

The interior is divided transversely into two unequal portions. The inner, which lies at a lower level than the other, is a dwelling one-storeyed and comprising a single chamber 51½ ft. by 19¼ ft., which contains two aumbries and narrow windows with spayed jambs to east and west. Chambers in the other part are represented only by mounds of debris, but it is probable that small angular chambers roofed with flags lay on either side of the entrance.

The earlier fort on Loch Caravat is marked as Dub Scor by the O.S. map on an islet in the northern part of the loch. No traces of defences are noticeable on this site, but on the eastern side is a small boat harbour running some 18 ft. into the island, with a width of some 6 ft. The adjoining island some 80 yards to the south-west, named Eilean Dubh Scor on the O.S. map, is apparently the fort itself.

More interesting to the naturalist or fisher is the fact that Loch Caravat is said, on good authority, to contain char, though there is no mention of the fact in any of the recognised authority, such as C.T. Regan. If there are char in Loch Caravat, probably or at least possibly they are to be found also in one or more of the lochs tributary to it. These are five in number, as we gather from the B.S.

Loch an Iasgaich is a little loch lying west from Loch Caravat and draining into it. It is fully half a mile long, is of the usual irregular outline and uneven bottom and is studded with small islands. A great part of it is less than 8 ft. deep and the maximum is 16 ft. The surface is about 11 ft. above the sea.

Loch 'ic Colla (McColl's Loch) is an extremely irregular loch a mile long. The south portion has an uneven bottom, the deepest sounding of 34 ft. having been taken near the east, while the north portion has a fairly deep basin at its west end where the maximum depth is 43 ft. Thus it is deep enough to hold char and certainly has good trout, averaging perhaps half a pound. A lesser basin 22 ft. deep lies to the east of this. The surface is 16 ft. above the sea. The stream, about 75 yards long, flowing into Loch Caravat, has a stony channel. In this channel the char probably spawn.

Loch an t'Seasgain is on the same level as the preceding and is really part of it; its maximum depth is 18 ft. and it is partly choked with weeds.

Loch na Ceithir-Eileana ('Four Islands'), the higher of the two lochs of this basin, which lies between Loch Oban nam Fiadh and Loch Obisary, is of

somewhat simple outline and is half a mile long.
The bottom is uneven and, as the name implies, four
islands rise above the surface. The maximum depth
is 42 ft. The height is 17 ft. above the sea. The range
of temperature on June 7, 1904, was great, the
greatest fall being observed between 20 and 25 ft:

Depth	°F.
Surface	64.6°
5 ft.	63.0°
20 ft.	58.0°
25 ft.	53.0°
35 ft.	52.0°

The fifth loch tributary to Loch Caravat is Loch na
Coinnich, a small loch half a mile long, between the
two last-named lochs, with a maximum depth of 25
ft. The height is 10 ft. above the sea. The range of
temperature here was also very high:

Depth	°F.
Surface	68.0°
10 ft.	63.0°
15 ft.	62.6°
20 ft.	57.0°
25 ft.	52.3°

Of these five lochs tributary to Loch Caravat, it is
plain that the fourth ('Four Islands') is most likely to

have char. It is the deepest and contains a considerable body of fairly cold water under 58°F. and above 52°F.

There remains one of the B.S. basins which contains two measured lochs, namely an upper and lower Loch a' Connachair, and an upper and lower Loch Hunder.

Loch a' Connachair, though draining through Loch Hunder into Loch Eport, is situated close to Loch Maddy. It is of the usual irregular form, and is shallow with a few deeper holes, with depths of 25 ft. in the south basin and 27 ft. (the maximum) in the northern. On June 3, 1904, there was a difference of 8.4° between the temperature at the surface and that at 25 ft., a fall of no less than 6.4° being observed between 15 and 20 ft. The loch is fully half a mile long.

Depth	°F.
Surface	60.4°
15 ft.	59.4°
20 ft.	53.0°
25 ft.	52.0°

In the Hunder basin, the lower loch is Loch Hunder, which drains into a branch of Loch Eport on its north side by a stream 40 or 50 yards in length. It lies on the west flank of the South Lee, and is 1¾ miles long by two thirds of a mile in greatest breadth. The outline is simpler than usual in the lochs of Uist. It is more like a valley loch but the presence of many islands indicates that it is not a simple basin. These islands divide the loch into three distinct basins. The northern basin is cut off from the middle basin by two large islands, the larger of which is joined by a causeway to the east shore. It has a maximum depth of 38 ft. The middle basin contains the maximum depth of the loch, 60 ft. It is separated from the southern basin by a chain of three islands. The middle one, called Dun Ban, is in the centre of the loch and has fairly deep water both to the east and the west. The eastern island is connected to the shore by a causeway, and on a smaller island is a large Dun. The southern basin has a depth of 55 ft., a short distance south of Dun Ban.

Another notable single loch in North Uist is Loch a' Bharpa which drains into the head of Loch Eport, between Loch nan Eun and Loch Tormasad. The western half of the loch, which is near a chambered cairn on the north slope of Ben Langass, is narrow and shallow, with several narrow inlets. The length is fully a mile, and the area over 30 ft. in depth occupies the centre of the loch, and is divided into two parts with maxima of 35 and 37 ft. The loch has good trout averaging half a pound.

Loch a' Buaille, on the north side of Loch Eport, between it and Loch Scadavay, is half a mile long,

but very narrow. It drains south through a smaller
loch into Loch Eport. The range of temperature on
June 4, 1904, is unusual:

Depth	°F.
Surface	67.0°
5 ft.	65.5°
10 ft.	62.5°
15 ft.	59.6°
20 ft.	52.5°

The height of the loch is 20 ft. above the sea, and
the maximum depth 14 ft.

Glas Loch nan Geireann (to distinguish it from the
salmon loch in the Scadavay basin) is a large loch in
the north-east of the island, two miles in length, with
a depth of 18 ft., and at a height of 16 ft. The short
stream conveying the overflow to the sea has a
rocky channel. There are many bays with white
sand, and a great part of the bottom is sandy, though
the shores are everywhere of rock, with some stony
stretches. The loch has ordinary trout and there are
some sea trout in it and in the sea pools.

Loch na Tomain is described in the lake survey as 'a
typical Uist loch'. It is nearly 1½ miles long, at a
height of 14 ft. above the sea and draining into the
Little Minch by a small stream one-third of a mile

long. It has four chief basins, the north eastern (the deepest) having a depth of 44 ft.

Loch an Duin, a tidal loch of exceedingly complex form, consisting of five principal expansions elongated from east to west, with many lesser inlets and numerous islands, is nearly a mile long and two-thirds of a mile in greatest breadth. All the inlets are shallow, mostly less than 7 ft. deep, except the northern, which has a hole where the depth is 35 ft. (known as 'Dead Man's Loch'). The two lowest basins connect separately with the sea and are filled with *Fucus*, *Zostera*, etc. The uppermost basin is slightly salt, and freshwater plants grow in it (e.g. *Myriophyllum*). The loch was 4 ft. above the sea on 20 May, 1904. There are two duns, as the name indicates, one in the northern branch, another, very well preserved, in the southern.

Loch Tarruin an Either is a loch of very irregular form, which lies between Loch Eport and Loch Scadavay, and drains by a stream 50 yards long into Oban Spanish, a branch of the former loch. It is half a mile in length and consists of a number of very narrow branches of little depth. Its height above the sea is 10 ft. and its maximum depth 23 ft. Its temperature on 4 June, 1904, had the great range of 14.5°.

Depth	°F.
Surface	66.7°
10 ft.	61.5°
15 ft.	58.2°
20 ft.	52.3°

Loch a' Buaille, on the north side of Loch Eport, between it and the southern extremity of Loch Scadavay, but exceedingly narrow, is shallow, the maximum depth being 23 ft. It drains south, through a smaller loch, into Loch Eport. The height of the loch is 20 ft., and, like the preceding loch on the same day, it had a range of 14.5° of temperature:

Depth	°F.
Surface	67.0°
10 ft.	52.5°
15 ft.	59.6°
20 ft.	52.5°

We may take next Loch Veiragvat, a small loch to the north of Loch Fada. It is 72 ft. above the sea, and the highest loch surveyed. There are several islands, the largest near the centre. The bottom is uneven and the greater part of it covered by less than 10 ft. of water, while the deepest place is a small hole between the island and the north shore.

Loch Tormasad is a sequence of two. The lower loch is two-thirds of a mile long and lies just west of the head of Loch Eport, into which it drains, through a smaller loch, by a burn half a mile in length. The bottom is nearly level, about 7 ft. deep, with depressions of 9 and 10 ft., but it holds only small trout. Loch Tormasad Beag, which runs into it, has, as we should expect, larger trout, but is shallow and rather peaty.

A neighbouring loch has much better fish. Loch Struben has trout averaging between a half and three quarters of a pound in its upper section, and still better trout - over a pound, and perhaps running up to 7 pounds - in its lower section.

Further north, on the west coast of the island, we come to the best trout lochs in North Uist. Near the village of Paible, we cross the Horisary River which comes from small lochs which have a large variety of sea trout – different from the usual sea trout of the island. In the rather scattered village of Paible, there is a fairly large, shallow, sandy loch, Loch Sandary, which is crossed by a road and holds some good trout of up to 4 lb. or so.

Further north, on the east side of the main road, we pass the small Loch Bhiorain, with trout averaging 2 lb., and running northward by a very small burn into the considerable Loch Eaval, which is fully a mile long and has trout averaging over half a pound. This loch is at a height of 28 ft. above the sea, and the outflow is used for a mill, running into Loch Grogary, near the extraordinarily primitive village of Hougharry, on the margin of the sea.

In the meantime we have passed two very good lochs – the 'Church' Loch, Loch nam Magarian, on the east side of the main road, and Loch Scarie on the west. Both have fine trout averaging a pound and a half or so. These lochs are only 7 ft. above the sea, and run into Loch Eaval and Loch Grogary respectively.

One other notable loch remains, Loch Hosta, which is given in the survey, and is about half a mile north of Loch Eaval. It differs from most of the other lochs in having a simple, little indented, outline. The basin is simple, the sides sloping gently all round, but a little more steeply on the north-east side, to the maximum of 31 ft. in the centre. The height of the loch is 23 ft. above the sea and a small burn, about a mile long, runs west to the sea at Raikinish.

Lewis

The island of Lewis and Harris is the largest of the lesser British Islands. It measures some 60 miles in length and 30 miles in breadth. Its southern part is mountainous, many peaks exceeding 2000 ft. in height, a few exceeding 2500 ft. The northern half is lower. The B.S. remarks of the lochs in Lewis:

There are many hundreds of lochs distributed over every part of the island. In the northern half they are especially numerous and in the central part they form a sort of watery maze like that of North Uist. There are only a few of the narrow, straight, valley lochs, so familiar on the mainland of Scotland, and these are in the southern mountainous part of the island; the majority are small, roundish, or relatively broad, and the larger ones of extremely irregular form.

To the majority of persons who are interested in ordinary trout, the island of Lewis is practically a *terra incognita*, and in the accounts of it, char are not even mentioned. There is indeed, in the N.S.A., the remarkable statement by the minister of the Parish of Lochs, that 'carp are to be met with here, but rarely' and, as usual, Stoddart faithfully copies the Account. But presumably 'carp' is merely a mistake for 'char', and the emended statement really conveys to us some information – that there are some lochs in the island that contain char – though we should have liked something more definite. The S.G. gives little help. It says that there are about 600 lochs in the island, but it gives merely a few lines to only six of them. The B.S., however, has measured thirty of the lochs and indicated their affiliations.

Five of the lochs exceed two miles in length. Loch Langavat is by far the longest, exceeding seven miles and in superficial area is about four times as great as any other loch. It is however, exceeded in volume by Loch Suainaval, which is also by far the deepest loch.

Loch Suinaval takes its name from a rather conspicuous hill (1403 ft.) and ought obviously to be called Suainavat as it is in the locality. It is 2⅔ miles in length, and at a height of 37 ft. above sea level.

Loch Suainaval is a simple basin, with the slope of the bottom steepest from the shore to the depth of 100 ft. The 100-foot contour closely follows the shoreline except at the ends, and the area enclosed by it is 2¼ miles in length. From the 100-foot contour to the centre, the slope is more gradual, and only two small areas exceed 200 ft. in depth. The larger of these areas, in the broadest part of the loch, is one-third of a mile long by one-fifth of a mile broad, and is flat-bottomed with a greatest depth of 212 ft. The lesser 200-foot area is a little south of the centre, is very narrow, and includes the maximum depth of 219 ft. A study of the contours shows that the loch has the U-shaped cross-section characteristic of lochs formed in valleys which have been occupied by glaciers (see Collet & Johnston 1906. *On the formation of certain lakes in the Highlands.* Proc. Roy. Soc. Edin., 26, 107). The mean depth, 108½ ft., is very great, more than three times that of any other loch in Lewis.

The volume of water shows that Loch Suainaval is the greatest loch in the island, though Loch Langavat is much longer (7¼ miles). The outflow northwards is via the short River Eyscleit, and eight or nine small lochs run into it, the largest about one third of a mile long. The temperatures observed on July 24, 1903, are interesting. They showed, consistent with the greater depth of the loch, a greater range than any other loch in Lewis.

Depth	°F.
Surface	57.0°
50 ft.	53.0°
75 ft.	50.4°
100 ft.	47.7°
200 ft.	45.8°

Loch Langavat is the second deepest loch in the island, with a maximum of 98 ft.

Loch Stacsavat is the lowest loch in the basin which contains Loch Suainaval. It is three quarters of a mile in length and is situated on the River Eyscleit which enters it from Loch Suainaval, while the River Forsa passes northward by a series of small waterfalls into the head of Camus Big. It is at a height of 36 ft. above the sea; the maximum depth is 40 ft. and the mean 17½ ft.

Temperatures on July 25, 1903, were as follows:

Depth	°F.
Surface	61.8°
20 ft	59.0°
30 ft	57.0°
37 ft.	56.4°

Considering the water temperature data derived from Loch Suainaval it is remarkable that the whole series from Loch Stacsavat should be so much higher than those taken in Loch Suainaval on the previous day.

Loch Raoinavat is a small loch about three miles north-east of Carloway on the west coast on a land surface rising gently towards the south. It is three-quarters of a mile long. It is narrow and at the east end expands and deepens westward. A very small area exceeding 50 ft. in depth, with a maximum of 61 ft. lies close to the north shore and near the west end. The mean depth is 20½ ft. and the drainage is barely half a square mile. The outflow is by a small stream which flows north past several mills into Loch na Muilne. The surface is 110 ft. above the sea.

Loch Grunavat is a fairly large loch lying about two miles west of Little Loch Roag, within a mile of Loch Suainaval. Though of the narrow form of valley lochs, it does not occupy a well-marked valley. High land occurs at the ends, while the sides are comparatively low. The length is 2¼ miles; the greatest breadth, about the middle of the loch, half a mile.

The shoreline is irregular, with several promontories and deep inlets. A large island occupies the exact centre of the loch and north of it is a simple and comparatively deep basin. The 25-foot and 50-foot contours extend a short distance south of the island, the deep channel passing between the island and the west shore. South of the island almost everywhere is shallow; halfway between the island and the north end is a small area over 75 ft. in depth, with a maximum of 90 ft. The mean depth is 28 ft. The area draining into the loch is three square miles; there are no inflowing streams of any size. The outlet, near the southern end of the loch, is through Loch na Ciste, which could not be entered by the B.S., is by the Gisla River, flowing two miles eastward into Little Loch Roag. The surface is 365 ft. above the sea.

Loch Raonasgail is a small loch among the hills which lie between the south end of Loch Suainaval and the west coast. It occupies a narrow valley between Tahaval (1688 ft.) on the east, and Mealisval (1883 ft.) on the west, which rise in steep crags covered with large and small stones on either side. The loch is of oblong form, two-thirds of a

mile long from north to south. It is a simple basin,
the slope of the bottom steepest on the east side, so
that the narrow area of over 75 ft. in depth lies near
the east shore. The maximum depth of 95 ft. lies
north of the centre. The mean depth is 32 ft. The
drainage area is two square miles. The Am Bealach
Raonasgail enters at the south end and the Amhuinn
Caslavat issuing from the north end, flows three
miles northward into the Camus Uig. The surface is
288 ft. high.

Water temperatures on July 29, 1903, were as
follows.

Depth	°F.
Surface	59.0°
25 ft.	56.9°
50 ft.	56.0°
93 ft.	53.0°

Loch Seaslavat is a small triangular loch close to the
shore of Camus Uig at Carnish, surrounded by low
hills. The length, from south-west to north-east is
over half a mile. The basin is simple, the bottom
sloping gently, except on the north-east side, which
is very steep, to the maximum depth of 82 ft. The
mean depth is 34½ ft. The drainage area is less than
half a square mile. The surface is 123 ft. above sea
level.

Loch Dibadale is a small narrow loch lying between Loch Suainaval and the marine Loch Resort. It is situated in a corrie between two hills, Mula Chaolartan and Tamanaisval (1530 ft.). It measures two-thirds of a mile from north-west to south-east. It is a simple basin and relatively deep, deepest in the southern half, with the maximum of 61 ft. somewhat south of the centre. The mean depth is 28 ft. The drainage area is 1¾ square miles. The Amhuinn Ghascleit flows out from the south end and joins the Tamanavay River. Loch Dibadale lies at the considerable elevation of 417 ft.

Loch na Crobhaig is a loch of moderate size, forming the lowest of the chain of lochs draining into the sea Loch Tamanavay, to the north of Loch Resort. The hills bounding its valley are much higher on the north side. On the flat southward towards Loch Bodavat are several small lochans. The length from east to west is a mile, and the greatest breadth, at the west end, nearly half a mile. The main loch is a simple basin, with the maximum depth, 50 ft., near the west shore. The mean depth is nearly 17 ft. It discharges by the Tamanavay River into the sea loch of the same name. The surface is 199 ft. above sea level.

Temperatures on August 4, 1903, were as follows.

Depth	°F.
Surface	59.0°
35 ft.	58.6°
40 ft.	58.0°
45 ft.	57.2°
49 ft.	55.2°

Loch Benisval is a broad sheet of water, half a mile north of Loch Resort. Low hills surround the loch, the highest being Benisval (621 ft.) to the south-east. The main loch is of oblong form, measured in a straight line from north to south. It is a simple and relatively deep basin, both sides steeply sloping along the base of Benisval, and the maximum depth, 95 ft., near the eastern shore. The northern end, where there are many small islands, is shallow. Only two lochs, Loch Langavat and Loch Suainaval, are deeper, and Loch Raonasgail is of the same depth. The mean depth of 35 ft. is exceeded only by Loch Suainaval and equalled by Loch Seaslavat. It receives only local drainage from an area of one square mile and discharges northward by the Amhuinn Benisval, a quarter of a mile long into Loch Criosdaig. The height is 276 ft. above sea level.

This list includes all the lochs on the island which have depths of 50 ft. or greater, with one exception, Loch Langavat, the largest loch with a maximum depth of 98 ft. It therefore includes all the lochs that are at all likely to contain char, and seems to give the completest answer possible to the problem suggested by the minister of the Parish of Lochs.

Hardie's manuscript stops abruptly at this point and it is unclear from his notes whether he perhaps intended to include a concluding paragraph or, indeed, a final chapter. He had succeeded in his major objective - a coverage of lochs and their salmonid fish in all parts of Scotland - and so perhaps no concluding remarks were needed. As a finale to both Hardie's volumes, we quote here parts of the objective review of Part I by W.L. Calderwood, Inspector of Salmon Fisheries of Scotland, published in the *Scottish Geographical Journal* in 1941.

> In spite of the title, this is a book about the lochs of Scotland. The author has consulted many works of early writers such as Stoddart and Colquhoun and Pennant, and has quoted freely from the Report of the Bathymetrical Survey for his descriptions of loch formations. ... there are particulars about many quite small lochs which are of interest ... in dealing with this or that district, the writings of local interest have been studied. ... although the author is evidently aware that ichthyologists now discard the name "Ferox" as in no sense a specific distinction, it is here continued as referring merely to large lake trout.

INDEX OF LOCHS

(Basins in parenthesis)

GENERAL INDEX